LAN Wiring

Titles of Related Interest

ISBN	AUTHOR	TITLE
0-07-005560-2	Black	*TCP/IP and Related Protocols, 2/e*
0-07-005590-4	Black	*Frame Relay Networks: Specifications and Implementation, 2/e*
0-07-020359-8	Feit	*SNMP: A Guide to Network Management*
0-07-021389-5	Feit	*TCP/IP: Architecture, Protocols and Implementation with IPv6 and IP Security, 2/e*
0-07-034249-0	Kessler	*ISDN: Concepts, Facilities, and Services, 3/e*
0-07-041051-8	Matusow	*SNA, APPN, HPR and TCP/IP Integration*
0-07-044362-9	Muller	*Network Planning Procurement and Management*
0-07-046380-8	Nemzow	*The Ethernet Management Guide, 3/e*
0-07-051506-9	Ranade and Sackett	*Introduction to SNA Networking, 2/e*
0-07-052134-4	Rhodes	*Building a Network*
0-07-057199-6	Saunders	*The McGraw-Hill High Speed LANs Handbook*
0-07-057639-4	Simonds	*Network Security: Data and Voice Communications*
0-07-069416-8	Summers	*ISDN Implementor's Guide*
0-07-063263-4	Taylor	*The McGraw-Hill Internetworking Handbook*
0-07-063301-0	Taylor	*McGraw-Hill Internetworking Command Reference*
0-07-063639-7	Terplan	*Effective Management of Local Area Networks, 2/e*
0-07-065766-1	Udupa	*Network Management System Essentials*

LAN Wiring

**An Illustrated Guide
to Network Cabling**

James Trulove

McGraw-Hill

New York   San Francisco   Washington, D.C.   Auckland   Bogotá
Caracas   Lisbon   London   Madrid   Mexico City   Milan
Montreal   New Delhi   San Juan   Singapore
Sydney   Tokyo   Toronto

Library of Congress Cataloging-in-Publication Data

Trulove, James.
   LAN wiring : an illustrated networking guide / James Trulove.
     p. cm.
   Includes bibliographical references and index.
   ISBN 0-07-065302-X
   1. Local area networks (Computer networks)—Equipment and
supplies. 2. Electric wiring. I. Title.
TK5105.7.T78 1997
004.6'8—dc21   96-51978
                                               CIP

McGraw-Hill

A Division of The McGraw·Hill Companies

     6 7 8 9 0 DOC/DOC 9 0 2 1 0 9 8 7

ISBN 0-07-065302-X

*The sponsoring editor for this book was John Wyzalek, the editing
supervisor was Paul R. Sobel, and the production supervisor was
Suzanne W. B. Rapcavage. It was set in New Century Schoolbook
by McGraw-Hill's Professional Book Group composition unit,
Hightstown, New Jersey.*

Printed and bound by R. R. Donnelley & Sons Company.

McGraw-Hill books are available at special quantity discounts to use as
premiums and sales promotions, or for use in corporate training
programs. For more information, please write to the Director of Special
Sales, McGraw-Hill, 11 West 19th Street, New York, NY 10011. Or
contact your local bookstore.

 This book is printed on recycled, acid-free paper containing a
minimum of 50% recycled de-inked fiber.

Contents

Part 3 Wire Management

Introduction

Knowledge is Power

One of the most challenging problems that you may encounter, when taking on a new project or a new area of responsibility, is finding factual, authoritative information on the subject. If you are new to the field of LANs and LAN wiring, you may be presented with a plethora of confusing, incomplete, self serving, and even misleading information on a variety of cable types and systems. You can find tons of information on every LAN topology imaginable and volumes of print on radical future technologies that are apparently "just around the corner". You need a reference that describes LAN wiring technology in detail to help you learn the basics and decode the market hype.

On the other hand, you may be an old hand at networks or the owner (or inheritor) of a mature LAN wiring system and be faced with important decisions regarding future upgrades to your network. What you need is a road map to LAN wiring technology, and you need an index to all the resources you will need to successfully complete your project.

This book will supply you with clear, concise information on the science and art of LAN wiring. LAN wiring certainly is a science, as much as any endeavor can be. Once you have picked a network topology, such as Ethernet or Token-Ring, and have chosen a wiring method, such as 10BaseT or UTP, the basic implementation is pretty much cut and dry. By default, you will have already defined the general wire type, minimum performance category, connectors, and even the connector wiring. You will know what maximum distances your cable runs may be, what wire size and specifications to use, and what standards to apply to your installation. Ah, but there is also much art added to the science.

The art of LAN wiring is in the realization that many factors must be taken into account to complete a successful installation. How well your particular installation approaches the exact-science ideal will be affected by such things as the details of workmanship, the influences of environmental factors, and even the quality of your wiring components. You must analyze your need for future enhancements and the potential of your particular choices to provide future compatibility with new technologies. In addition, cost factors must be weighed against their benefits. Your job must come in on time, under budget, and be done right. It must work properly with whatever level of performance was specified, and it must last for years with little or no maintenance.

The Four Sections

This book is divided into four main sections.

Part 1, "LAN Wiring Systems," describes the philosophies of LAN wiring systems, including the major LAN topologies. Structured cable plant design and planning are covered, along with such factors as mixing data and voice and general cost factors. The many types of cable used in LAN wiring are detailed, and the classic LAN wiring topologies and several popular universal cabling systems are described. A strong emphasis is placed on metallic wiring, as it is the dominant wiring method in current use. However, information on fiber optic wiring is also provided for those who are implementing fiber optic networks and for those applications, such as backbone and campus wiring, for which fiber is so well suited. The issues of future LAN wiring needs, migrating to higher speeds, and emerging technologies are also explored.

Part 2, "Wiring Devices," illustrates and explains the details of LAN wiring interconnection, from the telecommunications outlet to the wiring closet. Various types of wiring devices are shown for all phases of the wiring system. Charts are provided for the standard wiring patterns and common variations. Proper methods of installation and cable termination are covered, with emphasis on standards and workmanship. The total LAN wiring system also includes user cables, patch cords, and cross-connects, which are described as well.

Part 3, "Wire Management," discusses the planning and maintaining of LAN wiring systems. The location of wiring closets, interconnection, grounding, and power conditioning are covered in this section of the book. Approaches for cable testing, troubleshooting, certification, and maintenance are shown. In addition, physical wire system monitoring and control are described.

The last part of the book is comprised of the appendix, glossary, and index and provides reference information for the other three parts of the book. The appendix lists the various recognized standards for LAN wiring, including standards for cabling, telecommunications spaces, marking and identification, electrical safety, fire safety, and the common LAN topological standards, such as Ethernet and Token-Ring. LAN wiring is a constantly changing field and for that reason, we have also included some on-line resources to keep your knowledge up to date.

Who, What, and When

Who should read this book? *LAN Wiring* is written to be accessible as both a tutorial and a reference. This means that a LAN manager can learn the basics of connecting their LAN hardware, from design to implementation. The beginning LAN wiring technologist can learn the details of performing a proper LAN cabling installation. The expert will finally have a single reference to all the standards, wiring methods, and wiring details of LAN implementations. The manager will have a planning and estimating guide that clearly explains the terms, concepts, and details to specify in a contract with an outside vendor.

What information does this book provide? *LAN Wiring* covers planning, topologies, cabling systems, performance and safety standards, detailed connection and workmanship, and managing of the installed cable system. It primarily concentrates on twisted pair wiring systems, with some attention to fiber optic cables and legacy coax in the office environment.

When should you read this book? The information in *LAN Wiring* is useful at any stage of LAN implementation, from early planning to management. By all means, if you are planning to install, modify, or upgrade a LAN wiring system, you should use this book to assist you in the myriad of decisions, specifications, estimations, and evaluations for your network. Nothing is more important than the provision of a stable, reliable, and capable wiring system for your LAN. If you already have a mature LAN wiring system, you can use this book as a ready reference for standards, troubleshooting, and expansion. You will be easily able to determine what additional performance and/or speed your existing wiring is capable of offering. In addition, you can use the numerous lists, standards, tables, and illustrations to help decode manufacturer's literature and proposals.

Acknowledgments

The author would like to gratefully acknowledge the assistance of the following people in the preparation of this book: Donna Ballast, Alan Bookmiller, Jim Bordyn, Lee Haas, John Jankowski, Paul Kish, Henriecus Koeman, Mark Odie, Lisa Schwarz, Marjorie Spencer, and all the fine folks at McGraw-Hill, including, John Wyzalek, Paul Sobel, Roger Kasunic, Lori Flaherty, David McCandless, Midge Haramis, and all the artists, editors, production people, and others who helped in the creation of this book.

James Trulove

LAN Wiring
Systems

Design Philosophies
of LAN Wiring Systems

Local Area Network (LAN) wiring has developed into a highly sophisticated science with a tremendous impact on the performance, reliability, and maintainability of your network. While a LAN comprises many components, the underlying wiring system is the foundation upon which all else rests. Without the proper physical connections of copper wire, connectors, punchdown blocks, jumper cables, patch panels, user cables, and (perhaps) fiber optics, the network will not operate reliably.

It is therefore quite important that you properly plan, specify, and implement a LAN cabling system that will provide your organization with a reliable level of service. The use of the appropriate component parts and good workmanship in your wiring system will ensure that it can meet the performance standards that are expected for the type of LAN you are installing. In addition, you should always plan for as much future growth as is feasible, within obvious financial limits.

LAN wiring systems are in a constant state of change, as are all computer systems. As with computer technology, some of the changes in LAN wiring technology are evolutionary and some are revolutionary. For example, accepted LAN wiring techniques, components, and practices tend to *evolve* as manufacturers, installers, and users refine the existing wiring technology. The introduction of tighter cable link performance standards and the trend toward guarantees of component performance are perfect examples of evolutionary change. Likewise, the introduction of new outlet and patch panel designs is also evolutionary. You do not really see a major shift in an evolutionary new product; the refined product is just more convenient to use or has an enhanced level of performance.

On the other hand, some changes are revolutionary. A *revolutionary* change is one that breaks with the past. In technology, this implies a product or method that is no longer compatible with past items, but offers greatly increased capability. For example, an increase in LAN speed is revolutionary. For the most part, the technology required to increase from 10 Mbps to 100 Mbps is a clear break from the previous practice, and is highly desirable, but no longer compatible with the past. Revolutionary changes tend to be abrupt.

The main focus of a LAN wiring system design effort must be to keep up with evolutionary changes in the wiring technology that will enhance the value of your system, while attempting to plan for the next revolutionary jump in LAN technology.

When LAN wiring over twisted-pair wire first began, little was known about the importance of wire performance. The initial interest in "twisted pair Ethernet" was to be able to use one's existing telephone wiring for the installation of a network, thus saving the enormous cost of installing new cabling. However, we soon found out that the typical twist-per-foot cable specifications that were fine for telephone use were rather marginal for LAN frequencies. To make matters worse, deregulation of the telephone industry meant that cost-conscious installers sometimes used a cheap grade of telephone cable that had two twists per foot or less, making it inadequate for LAN data use. To use twisted pair with confidence, the LAN manager and the LAN installer needed to be certain that a particular cable would perform adequately. So, a performance specification began to evolve that categorized cable performance by intended use. An early division of cable manufacturing quality levels eventually led to a well-researched and detailed set of performance specifications in five categories. The LAN user now could know with certainty that a Category 3 cable will operate well on LANs up to 16 megahertz (MHz) bandwidth, and that a Category 5 cable will perform to 100 MHz.

In this way, many connector and wire specifications have evolved, along with workmanship practices, to provide a platform of standards that virtually guarantee a properly performing LAN wiring system. This process is not difficult to understand, but the LAN manager, installer, and technician must have a thorough understanding of the standards, procedures, and specifications need for a network cabling plan that complies with LAN operating standards. The careful planning of a standards-compliant cabling system is called *structured wiring*.

This chapter will give you an overview of the philosophies in designing, planning, and estimating a LAN wiring system. The items in these sections should be ready knowledge before you delve into the rest of the book. A central theme of this book is that any wiring design should be a standard installation that supports at least the minimum performance criteria, is simple to install and maintain, and provides adequate future expandability and growth.

A Standard Installation

Maintaining the proper standards in your cable installation is the key to getting the most from your LAN cabling system. The internationally recognized standards exist to ensure that the combination of wiring, connectors, hubs, and network adapters will all perform properly in a completed network. The newly revised wiring standards make special note of specific cable installation guidelines and workmanship practices that have been found necessary for the more demanding modern networks at speeds of 100 Mbps and beyond.

There are standards that apply to LAN wiring covering everything from electrical performance to safety issues. Should you dream up an exotic wiring system that meets your creative needs? Should you try to use some existing telephone wiring and add whatever else you need to make the system work? Should you pick and choose from various manufacturers' catalogs? How do you sort out all the options?

One way to proceed is to choose a very conventional LAN wiring installation that uses the standards to your advantage. In such an installation, you will wire your cables in a manner that will support the widest variety of current and future applications. This means you must use the proper cable, install it properly, maintain lengths within the maximum proscribed distances, use only connectors and jacks that meet the category of operation you need, and use proper installation techniques. In addition, you will want to thoroughly test and document your cable system for the appropriate level of performance. Of course, any future changes and additions to your standard cable system must also be done the same way.

Designing a standards-compliant, conventional installation will have distinct advantages for you. For example, your properly installed cable will never contribute to a connectivity failure. You can be assured that any problems will be found in the network hardware or software that utilizes your cabling system, not in the cable itself. That is true, assuming no inadvertent change has occurred in a cable. Naturally, it is always possible that someone could damage a cable while doing unrelated work on an electrical, plumbing, HVAC, or even telephone problem. Remember also that user cables (from the workstation to the wall jack) are considered part of the cabling system, and their failure can indeed disable a network connection.

Your standard installation will also ensure that any trained cable installer will be able to easily expand, troubleshoot, or repair your cabling system. If you test your cable when it is installed, you will be able to retest any cable drop and compare the results to the original cable certification done at installation. You will also know that your cable system meets the electrical and fire protection standards required by local and national authorities.

The standards you will need to reference are described throughout this book and are listed in the appendix. The appendix also contains sources for copies of the pertinent standards.

Installability

Your LAN wiring system should be easy to install and maintain. A variety of wire types, connector types, termination devices, and patching devices exist to support network wiring. The best choices are those that: support your applications, present and future; and are easy to properly install, using a minimum of special equipment.

However, you must make many additional choices concerning the particular wiring components to be used. For example, what types of jacks are the easiest and most reliable to use? Should you use a punch-down termination and cross-connect wiring in the wiring closet or should you terminate cables directly onto a patch panel? Where should the wiring closet(s) be located? How should the individual cables be run to avoid electrical interference and other performance decreasing factors? What color should the cable and connector plates be?

Some guidance to these choices is provided by recognized installation standards, but many are a matter of personal preference. For example, some jacks use wire termination methods that require special tools, while others do not. The experience level of those who will install and maintain the cabling system is also a factor. Some items are more easily installed and use such things as color code marking to simplify the process. Outlet plates are available that have modular construction, allowing you to mix connections for data, phone, and even CATV.

Installability will also have an effect on system cost. Whether you are doing the wiring yourself or using a contractor, factors such as time-to-install and reliability will be important. The easier the wiring components are to install, the lower will be your cost of installation, and a contractor will charge less because the installation will take less time. Doing the work yourself will cost less as well. Installability, however, does not mean the taking of shortcuts. Your goal is to achieve a very installable cabling system that is easy to maintain and meets all appropriate standards and performance criteria for your present and foreseeable future needs.

The sections on wiring devices in Part II of this book will show you a variety of wire termination devices, such as jacks and patches, and will compare the different installation techniques that impact installability.

Reusability

Everybody involved with computers and networking knows by now that technology moves very quickly. Changes are constantly made to upgrade performance and throughput. (As we all know, nothing stays the same but change itself.)

You cannot always afford to install the true leading-edge technology, but you should at least design your cabling network to allow for a future upgrade without replacing the cable. To do this means that you must design a cabling system that totally meets your current needs. If the current need is 10BaseT, then the cable system should be specified so that it meets that requirement, as a minimum. If possible, it may be convenient to design your system so that it could support multiple applications, such as data and voice, or data and video conferencing.

Then you should assess what future standards for which you might reasonably be expected to use the cabling system. You should also take into account the useful life of your facility, which might be five years, or even 20 years. Are your offices in leased space? You might expect to move in less time than if you owned the space. It might be best to "over-wire" by putting in more cables or cable pairs than you currently need. You may wish to use a modular outlet jack that allows the connector to be changed without reterminating the wire.

It may be virtually impossible to make any accurate technological prediction more than five or 10 years out. Consider for a moment what technology changes have taken place over the last 10 years. Undoubtedly, someone, somewhere made an accurate prediction. However, there were also several predicted technology directions that simply didn't come true. Of the predictions for the next decade, which should you choose?

The only practical approach to cabling futures is to determine a cabling method that meets at least the current, widely deployed technology requirements. You then should consider installing to meet cabling standards that are at least one level beyond those current requirements. If this is not cost-prohibitive, your cable system will be able to go through at least one generational upgrade without replacement. For example, if you currently need a 10BaseT or 16 megabit (Mbit) Token-Ring network cabling system, you should consider installing a cabling system that will support 100 Mbit technology. Since much of the current componentry is being manufactured so as to exceed that data rate, your system will probably support a second upgrade to 155 Mbit rates, should that occur.

Changes in cabling technology are a challenge for any book author. To help you keep informed on the latest developments and standards, we have developed an Internet site to use as a resource on current LAN developments. See "Online Resources" in the appendix for more information.

Reliability and Maintainability

Does anyone ever want an unreliable network cabling system? No, of course not! You, or someone else if you use a third party installer, will be making many decisions regarding the components and methods used to install your LAN cable. Many of these decisions will have a long-term effect on how well your network performs.

On the other hand, if you install cable for someone else, you will want to make sure that the installation will stand the test of time. Your reputation is on the line, and you may be responsible for repairs should failures occur at some later time.

A very good philosophy is to install LAN cabling in such a manner that, when network problems occur, it will never, *never* be the cable system that is at fault. If you adhere to this guideline, you will save countless hours of tedious and expensive troubleshooting. Even though wiring components are relatively simple physical devices, they have technically sophisticated operating characteristics that require expensive equipment and trained technicians to troubleshoot. The performance issues are greatly compounded if you use the wiring to its maximum capability.

Because many companies with LAN installations do not have access to sophisticated wire test equipment, many LAN managers start troubleshooting a problem by looking at the servers, routers, hubs, and workstations connected to the cabling system. They check the configurations of network software, drivers, and applications. They may even change out hardware, including computers, printers, hub cards, or whatever seems to be related to the problem. After all that thrashing about, you never want the cable to be at fault. Too much emotional and physical energy has been tied up in troubleshooting by the time they get to the cable. It had better not be a cable fault!

So, to prevent all of this energy from getting expended at you, just make certain the cable system is done right. Proper installation of LAN wiring has become considerably more difficult as LAN speeds push upward. You should install all your wire in accordance with the latest standards, particularly those that address cable routing, wire center locations, handling of the cable itself, and workmanship. Proper planning, components, installation practices, and workmanship will make your cable system installation both reliable and maintainable.

The Structure of Cabling

Basic to the philosophy of modern LAN wiring is the concept of structured cabling. With the higher speeds of today's networks, it is recognized that the total length of cable that connects from the hub to a workstation or other device has a finite maximum length. The entire networking system must therefore be broken up into chunks that allow workstation (or station) wire to be concentrated, with each cable length short enough to support the desired data rate. Structured wiring standards have been developed to help the LAN user plan a wiring system that stays within the maximum wiring distance for various LAN topologies. For example, in the case of 10BaseT, this cable must be no more than 100 m (328 ft), including patch and equipment cords.

We achieve the needed wiring concentration by placing wire closets (wiring centers) at appropriate locations in a building and then interconnecting those wiring closets as needed to provide the total network connectivity for the building. Typically, a model of a multistory building is used to illustrate this structured concept, as shown in Fig. 1.1.

On each floor of our model, a central wire center concentrates all of the station cables for that floor. Each workstation location has a wall or surface-mounted jack. The network cable is terminated in that jack and runs directly to the wire center. This is called a *home run*, as there are no intermediate connections, taps, or daisy-chains. The wire may run in wire trays or conduit, be draped over supports, or run over a drop ceiling (the type with push-up ceiling tiles). Larger floors may require more than one wire center per floor.

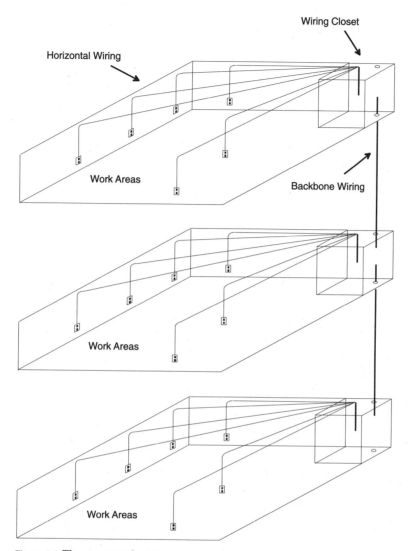

Figure 1.1 The structured wiring system.

At the wire center, each station wire is terminated on an appropriate punch-down termination, or directly onto a patch panel location. The punchdowns or patch panels may be mounted to a wall or in a freestanding rack or cabinet. In the wire center, some type of network device, such as a hub or concentrator, is connected to each station cable and electrically terminates the cable run. The hub passes the LAN signals on to other stations or wire centers for ultimate connectivity with the entire network. The process is essentially the same for different network topologies, although Token-Ring uses passive wiring concentrators called MultiStation Access Units (MSAUs), rather than the active hubs of 10BaseT.

A wire center is typically connected to the wire centers on other floors. This center-to-center wiring is usually done from floor to floor to floor, as a backbone, with LAN hubs on each floor. In some cases, it may be more effective to connect wire centers on several floors to a single backbone concentrator on one of the floors. Ideally, wire centers should be located directly above one another, to minimize the cable runs between them, but that varies from building to building. Fiber optic cable is sometimes a good choice for wiring between wire centers, as it totally eliminates grounding and bonding concerns that exist with metallic cable and can often be run longer distances than copper cable.

Obviously, there are differences between the wiring considerations for station wiring on a single floor and for center-to-center wiring between floors. For one thing, the fire protection requirements may be different for the vertical riser cabling used between floors. Physical securing of the riser cable may also be different, as may the need for strength. For this reason, it is convenient to have terms for the two wiring paths that easily distinguish the concepts and physical characteristics of the cable.

Station wire from the wire center to the workstation termination is simply called *horizontal wiring*, a somewhat logical name as most of the wire run is horizontal across a floor of a building. The center-to-center cables are called *backbone wiring*, since they form a unifying structure between wiring closets. Of course, a large floor of a building would have more than one wiring closet, to adhere to the station wire maximums, but the wire between those wire centers would still be referred to as backbone. All the details of this structured wiring concept are covered in the next chapter.

Cost Factors in Cabling Systems

A variety of methods exist to implement LAN wiring schemes; most methods in common use will be described in succeeding chapters. These methods vary in three significant degrees, all of which have cost implications.

First, some LAN topologies require that certain types of wire and connections be used. For example, the Ethernet 10Base2 topology requires coax, BNC T's, and 50-ohms terminators. Normal 10BaseT can use Category 3 twisted pair and jacks, while 100BaseTX requires Category 5 twisted pair wire and jacks. Although your network requirements will frequently dictate which topology you must use, you may have several choices to make that will influence initial cabling cost as well as the cost of future expansion. In some cases, it would be possible to choose an older wiring technology such as 10Base2 coax on the basis of cost. However, when you consider in the cost of a future upgrade to 100 megabits-per-second (Mbps), a Category 5 unshielded twisted pair installation would support 10 Mbps 10BaseT today and 100 Mbps 100BaseTX for the future. Second, LAN wiring methods vary in the extent to which they will support higher speed network data rates, including future standards that may not be fully implemented yet. For example, it is less expensive to install a Category 3 cabling system that supports your currently needed 10 Mbps 10BaseT or 16 Mbps Token-Ring network than it would be to install to a Category 5 facility. You might never need the 100 Mbps capability, particularly if your facility will be occupied for less than five years. What if you had to stay longer? Should you put in Category 3 jacks and cable to save money, or should you go ahead and install a Category 5 system now, in case the future comes along a little sooner than expected?

Third, you may choose combinations of cable and connection hardware that differ in features and cost. This factor may also influence the final, installed cost of your network cable system. Even among manufacturers that offer products certified to identical performance ratings, there is a lot of price variation. You may choose a big-name manufacturer whose wire or terminations cost more than the little guy. Some manufacturers have products with extra features, such as modular snap-in construction or fancy color inserts. Some even offer special 10- or 15-year warranties if you use their cable, connectors, and certified installers. Others offer separate wire termination and jack modules that plug together. In many cases, the extra features will cost you extra money. You will have to judge the benefit of these extra cost features for your situation.

In the other sections of this book, we will try to point out the cost tradeoffs of the devices that are described. You should be aware that anytime a connector, cable, or other wiring device is referred to as "easier to use" or "flexible," there might be cost implications. Remember too that your goal is not necessarily to specify or install the cheapest wiring solution, but to find the balance among features, cost, and reliability that meets both your current and planned future needs.

Mixing Data and Telephone

Combining telephone and data on the same cable is sometimes a consideration for cost savings and efficiency. This so-called mixed media approach is an option that may work fine in a lower-speed network environment, such as 10BaseT or Token-Ring, but it may be fraught with problems at higher LAN speeds. (At the least, there are lots of "ifs.")

If your telephones are analog, your LAN speed is conventional, and the telephone cross-connect is in the same wiring closet as the LAN hubs, then you probably could use the same cable for data and phone very successfully. This case assumes that you are using at least Category 3 Cable, the minimum data grade, and that the cable runs are not too close to the 90 m maximum.

However, if your telephones are digital, or you intend to use 100 Mbps LAN data, or you are installing the LAN hubs in a separate closet from the telephone interconnects, then you should run two separate cables for data and voice. For that matter, if you know that the telephone installers are apt to be

TABLE 1.1   Cabling Recommendations for Various Applications

Application	Minimum cable performance categories		
	Cat. 2	Cat. 3***	Cat. 5
Analog phone only	Yes	Yes	Yes
LAN 4/10/16 Mbps only	Yes (for runs under 100 ft.)	Yes (for runs under 290 ft.)	Yes (for runs under 350 ft.)
Shared analog phone and LAN 4/10/16 Mbps	Yes* (if 4 pair cable)	Yes* (if 4 pair cable)	Yes* (if 4 pair cable)
Digital phone only	Not recommended	Yes	Yes
Shared digital phone and LAN 4/10/16 Mbps	No	Not recommended	Not recommended
LAN 100 Mbps only	No	Yes/no**	Yes
Shared digital phone and LAN 100 Mbps	No	No	No

*May not operate at extended distances, particularly with Token-Ring 16 Mbps networks. Higher performance category cables will provide better isolation between the phone and data signals.

**Some 100 Mbps technologies are designed to operate on a minimum of Category 3 cable.

***Category 4 cable may be substituted for Category 3 applications. In general, any higher performance category may be substituted for any lower category cable.

overly creative around the LAN wiring, you should run separate cables (color-coded for added protection). Table 1.1 shows some of the typical recommendations for telephone and data.

The reason for the concern for mixing voice and data is the phenomenon of crosstalk that occurs between pairs of the same cable. The higher frequency components of 100 Mbps LAN signals, along with their greater attenuation per foot of cable run, make them much more susceptible to interference from other sources. Such sources clearly include digital telephones and other LAN signals. Even magnetic fields from fluorescent light fixtures can cause problems. At cable run distances that approach the maximum limit, these LAN signals are significantly more susceptible to problems from crosstalk, signal balance, interfering signals, magnetic coupling, and noise sources.

The danger of interference from telephone pairs is probably less for the analog signals of conventional telephones, since they limit transmitted frequencies to 5 kHz or less. However, both analog and digital telephones may use extra wire pairs for power. This power connection is unbalanced and may serve to increase the undesired coupling effects between pairs in the cable.

Some cabling designers feel that data and telephone wiring should never be mixed. There is some support for this view from the standards, when interpreted strictly. None of the standards provide guidelines for the application of two or more signals to the same cable, although some users simply ignore the issue. The most widely used standard, TIA 568-A, simply suggests two cables to each telecommunications outlet and recommends that each be used for only one application. If you are going to provide separate cables for phone and data, you can easily run them to totally separate areas of a wiring closet, or even to separate wiring closets.

You will never cause a technical problem with either LAN or telephone systems by separating data and voice cables. On the other hand, you can expect the installation cost to be higher in almost all cases.

2

Wiring System Design

The three keys to a successful LAN wiring installation are proper design, quality materials, and good workmanship. This chapter will cover the basics of proper wiring system design. Proper design involves the careful orchestration of several complex factors, applied in a standard fashion, to produce a successful installation plan that will meet your needs for today and for many years to come.

These factors include the length of run, wire type, wire terminations, and routing. Many of the technical details of these factors are covered in other areas of this book. In this chapter, we will give you an overview of how these factors come into play in creating a successful design.

In creating your cabling design, you will have to make many decisions. Most of these decisions can be reduced to a simple matrix, so that your overall performance requirements actually dictate the proper components to choose. In this way, you will insure that the final wiring system design will provide the connectivity, performance, and growth that you need.

This chapter gives a brief explanation of structured cabling concepts with a summary of some of the design considerations involved and culminates in a simple design example.

In later chapters, we will explain many of the common wire types and wiring systems that are, or have been, used in LAN wiring. Although some of the older cabling methods are no longer being used in new installations, it is important that they be here for completeness and for those who must add to or modify older cabling systems. For the most part, we will talk about unshielded twisted pair wiring when citing design factors and installation techniques. This has been the area of heated activity in the specification of wire, installation and performance standards, and introduction of new technologies. This is clearly where the action is.

Fiber optic cabling is also a current technology, but its use in the workplace is limited at present. It is more expensive to buy, more difficult and expensive to install, requires more expensive workstation and hub interfaces, and generally exceeds the bandwidth required for the current and next generation of LAN data throughput. Fiber optic cabling does, however, have several unique and very beneficial characteristics that can be of great assistance in larger cabling designs. For that reason, a description of fiber optic cabling is included wherever that medium is particularly useful.

Basic Terminology

In order to talk about the intricacies of a subject such as LAN wiring, we need a common lexicon. Unfortunately, the cabling industry has developed as a combination of many technologies and disciplines, each with its own vocabulary and terminology. Many of the twisted pair wiring techniques we use in today's LAN wiring were initially the domain of the telephone industry. In fact, the styles of wire, connectors, terminations, and even color codes that we use today were developed decades before anyone considered using telephone-type wire for data. Many of these telephone terms have been made directly a part of LAN wiring vocabularies. On the other hand, all of the terms that are used for LAN wiring signals, cables, standards, and termination equipment come to us from the computer networking industry. Now that we are "pushing the envelope" into the stratosphere of LAN speed performance, we are beginning to encounter even newer terms that may be unfamiliar to many of us, but would make any radio-frequency (RF) engineer smile.

In this book, we will attempt to be as consistent as possible in terminology so that you will always know what we mean by a particular word or phrase. We will particularly avoid jargon that is used only by a narrow segment of the industry, other than to explain special meanings of terms that you may encounter when dealing with contractors and suppliers. It will be necessary to use some terms almost interchangeably, such as the terms wiring and cabling. To not do so would be a disservice to you the reader, since you will certainly encounter all of these interchangeable terms in your everyday work. We will always try to be clear in our meaning.

A complete glossary appears at the end of this book. Each time we introduce a new term we will define it. Some LAN wiring terms are common words or phrases that have a particular meaning in the cabling industry. For the most part, after these are introduced, we will simply use them with their special meanings, without emphasis. In a few instances, special words or terms will be capitalized. Two examples of this are Basic Link and Channel, which are very specific terms used in one of the service bulletins that revise, amplify, or clarify EIA/TIA standards. Such terms are not in general use and their meanings are specific to the definitions in a standards document. Anacronyms and abbreviations are also part of the world of LAN wiring. The common terms will be formally introduced only once, although the less common ones may have their definitions repeated at the beginning of a section.

Specific, commonly accepted terms can be very useful in providing shortcuts to understanding and explanation of a complex system such as LAN wiring. However, jargon that is peculiar to a narrow part of an industry can be tedious and even confusing. We will therefore stick to more commonly used computer networking terms that are widely understood. If you have a technical background in a specific area, such as the telephone industry, you should easily understand the terms we use. Please do not be offended if we refer to LAN wire to the workstation as UTP cable, or horizontal cable, rather than IW (inside wire) or station wire. The former terms are much more specific to the LAN wiring that is our subject.

Twisted Pair Wiring

The basic wire type for most current LAN wiring installations is called *twisted pair*. Some refer to this wire as "unshielded" twisted pair (UTP) to differentiate it from "shielded" twisted pair. This style of wire is relatively inexpensive, is easy to connect, and provides self-shielding properties to minimize harmful interference to or from the cable. Twisted pair wire was originally used by the telephone industry to enable the widely separated individual wires that once populated telephone poles to be replaced by wire pairs in very close proximity. This twisted pairing allows more than one telephone conversation to be carried within the same cable jacket. The twisting causes the electromagnetic coupling from one wire to another to be canceled out and eliminates interfering crosstalk between pairs. Any significant amount of crosstalk coupling limits the length of cable that can be used. While the amount of coupling is insignificant at audio (voice) frequencies, it becomes much greater at the higher LAN frequencies. That is one of the reasons we do have to limit the length of LAN cabling. The construction of twisted pair cable is described in more detail in Chap. 3.

The use of twisted pair cable is a key component in LAN wiring system design. The properties of this type of cable define the maximum usable length of a workstation cable at any given LAN data rate. The characteristics of LAN cable are carefully specified in the applicable standards. Particularly at the higher LAN speeds, these performance characteristics are the limiting factor. At 100 Mbps, we are pushing the capabilities of twisted pair wire. At 155 Mbps, we may be at the practical limit for a useful cable length to the workstation for two pair connections, but several new methods use all four pairs and are still not at the limit.

The Electronic Industries Association (EIA) and Telecommunications Industries Association (TIA) standards specify the maximum length from the wiring closet to the workstation as 100 m (90 m for the horizontal cable, or station cable to the wall jack, and 10 meters for the patch cord plus the user cords). This is approximately 328 ft, and should be sufficient for most building installations. However, it does put one at the outer limit of performance where everything becomes critical, from routing and handling, to measurement and certification.

For example, several proposals of the ATM (Asynchronous Transmission Mode) Forum specify twisted pair wiring for the medium. These proposed new connectivity options range in speed from 25 Mbps to 155 Mbps and use the TIA 568-A 90 m (295 ft) horizontal cable link as the model. Some high-speed networking proposals use two pairs, while others use four pairs. The higher speeds may require the use of premium Category 5 cable and connecting hardware, but they still operate on conventional metallic twisted pair cable. Clearly, twisted pair wiring has a lot to offer for the future of networking.

Structured Wiring

Structured wiring based on the standards (such as TIA 568-A) assumes that the total wiring system will be divided into simple wiring units. These wiring units can be repeated as needed, combined into larger structures, and those structures interconnected to product the overall wiring system. Although some standards exist for implementing particular LAN topologies, such as 10BaseT, over a wiring system, most of the standards simply require a generic, multi-purpose, reusable wiring and cabling system that could be used for anything from voice to video. Of course, these standards are designed to allow LAN traffic to properly operate at specific data rates, and specify test parameters to insure performance. However, the standards for the cabling itself are not at all specific to any type of LAN.

This is an enormous advantage to you, as the cable system designer, installer, or LAN manager. If you install a standards-based structured wiring system, you do not need to know anything at all about the exact type of LAN that will be installed in your wiring system. All you do need to know is what data rate will be expected of the LAN, because that will define the category of cable that must be used.

The most significant standard in LAN wiring is the "TIA/EIA-568-A 1995 Commercial Building Telecommunications Wiring Standard." Along with related standards, and bulletins, this standard defines a universal cabling system that meets a variety of needs. We will be referring to this standard often in this book, and so we will use an appropriately shortened version of the complete name, "TIA 568-A," which is easier to read and even easier to say aloud. This is a permissible nickname, since the TIA is really the moving force behind this standard, although it is approved by EIA and ANSI and coordinated with many national and international standards. (By the way, one normally says the letters T, I, A individually, although you could certainly say it as a word.)

Any type of voice, data, or LAN signals can be connected to your structured cabling system. For example, the twisted pair wiring system that is specified by TIA 568-A will allow you to run analog or digital voice, 10BaseT Ethernet, 100BaseTX Ethernet, 100BaseT4 Ethernet, 100VG-AnyLAN, LocalTalk / AppleTalk (tm), Arcnet, 4 Mbps Token-Ring, 16 Mbps Token-Ring, 100 Mbps Token-Ring, CDDI (Copper FDDI at 100 Mbps), and ATM at 25, 51, and 155 Mbps. Of course, ISDN (Integrated Services Digital Network), Switched

56 kbps, T1 or E1, and many other telephone services are also compatible with structured wiring systems. Also, several products to distribute video over twisted pair in NTSC format have been introduced, and you can even distribute IBM-5250 and -3270 terminals over the same wiring system.

Such a wide range of uses makes the term generic seem totally inadequate. Perhaps multipurpose or general use would better describe this structured cabling system.

The true versatility of the TIA 568-A wiring system is a powerful advantage to any organization that uses it. For this reason, you should be wary of compromises that limit your wiring system's utility. Although you may have a very specific purpose in mind, who can say to what use the wiring system will be put over its useful life of 10 or 20 years?

International standards also cover much of the same ground. For the most part, these standards are all coordinated. It would be tiresome indeed to constantly list all of the coordinated or referenced standards for each citing. Therefore, we will simply use the U.S. standard number in most cases. A fairly complete list of U.S., Canadian, European Community, and International standards is given in the appendix of this book. Standards that are coordinated are so noted.

The structural wiring system comprises several component structures. Within a building, the structures are called horizontal and backbone wiring.

Horizontal wiring structure

The basic component of the structured wiring system is the *horizontal wiring* (cabling) structure. Horizontal cabling consists of the cable and connections from the wall or other jack at the user workstation outlet to the wire termination in the wiring closet. As shown in Fig. 2.1, the horizontal wire may include a punch-down block, cross-connect wire, and a patch panel. However, it does not include hubs, routers, or network adapters. Horizontal cable is defined in TIA 568-A, "Commercial Building Telecomunications Wiring Standard," and is the Basic Link (minus cross-connect and test cables) defined in TSB-67, "Transmission Performance Specifications for Field Testing of Unshielded Twisted-Pair Cabling Systems." In practice, horizontal wire is routed directly from the wiring closet to the workstation, without any intermediate splice points, cable junctures, or taps, as might be common practice in traditional telephone wiring. This direct-wiring technique is referred to as a *home run*.

Various wiring standards put special qualifications on horizontal wiring. Some of these may limit the types of components, such as cross-connects, that may be a part of horizontal wiring. More specifically, you should make sure that you use cable and components that are rated for the data rates you intend to use. For example, if you are installing a Category 5 (TIA 568-A) facilities, you must use Category 5 rated jacks, cable, patches, and user cords. If you use a cross-connect field to connect between the station cables and a patch panel, the cross-connect wire must also be rated Category 5.

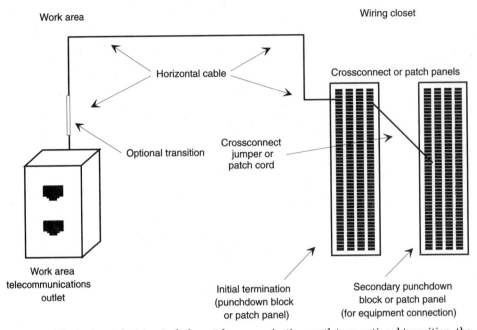

Figure 2.1 The horizontal wiring includes a telecommunications outlet, an optional transition, the horizontal cable, and alternatively a punchdown block and crossconnect wire, or a patch panel and patch cord.

In the horizontal wiring structure, the station cable is terminated in the area of each user workstation with an appropriate jack. The jack is typically an 8-pin modular female connector in a flush-mount wall plate that mounts in a single-gang utility outlet box or metal attachment ring secured in a hollow sheetrock wall. The 8-pin modular jack is sometimes referred to as an "RJ-45," because the connector/jack components are the same. However, RJ-45 actually applies to a special purpose jack configuration that is not used in LAN or standard telephone wiring. Several types of 8-pin jacks are available, and may be single (simplex) jacks, dual (duplex) jacks, or modular snap-in jacks of up to 6 jacks per plate. Solid walls, plaster walls, or hard-to-reach locations may use a surface mounted jack enclosure (such as a "biscuit block," or a flush-mount jack in a surface mount box) and appropriate surface raceway to reach the user outlet location.

Modular furniture requires special consideration, because the workstation cable may need to be dropped from the ceiling or run from an adjacent wall into wire channels within the modular furniture units. Some modular furniture provides special channels for telephone and computer wiring that is separated from the AC power wiring channels. Caution should be used when sharing the power wiring channel with data cabling. Not only is there a potential safety concern from having the LAN wire so close to power wire voltages, but there may be performance problems, particularly for Category 5 performance. Some guidelines specify a 2-in or more separation from power wires, while others specify a greater distance. In addition, local and national installation codes generally prohibit low-voltage wiring from being installed in the same conduit as power wires. Some local authorities may consider the open wiring channels of modular furniture to be in fact a conduit and therefore subject to this rule.

Horizontal wiring is specified according to performance categories defined in TIA 568-A. Both cable and connecting hardware (jacks and plugs) must meet performance guidelines appropriate for their use. Not all cable or components meet the minimum standards required for LAN use. One of the factors affecting performance in outlet jacks is the method of wire termination. The station cable connects to the jack by means of either screw terminals or some sort of insulation-displacement termination. Screw terminations are acceptable in Category 3 installations, but are generally prohibited in Category 4 or 5 installations, due to performance constraints. Category 3 is appropriate for 10 or 16 MHz bandwidth networks, Category 4 for 20 MHz, and Category 5 for 100 MHz. For the most part, there are no networks that require Category 4 specifications, although it may be used if you desire a "high-grade" Category 3 installation. Some references suggest Category 4 for Token-Ring networks, but the current standards require only Category 3.

Insulation-displacement terminations use a split metal contact upon which the insulated wire is forced, thereby cutting through the insulation and making metal-to-metal contact with the copper wire. Several types of insulation-displacement connectors (IDCs) exist, including modular snap-together IDC jacks, modular 110-type IDC jacks, and older 66-type IDC jack plates. Except for the 66-type jacks, all are commonly available in models that are certified to Category 5 performance standards. Category 5 66-blocks do exist, but the installer may have difficulty maintaining the required twist all the way to the termination point.

From the workstation outlet jack, the horizontal station cable is run to the designated wiring closet. This cable run is designed to be within (possibly well within) the 90-m maximum specified by the standards. It may be routed with other individual station cables in wiring trays, along ceiling tiles, over support beams or frames, through firewalls, or even in conduit. If you have special performance criteria, such as Category 5, you should take precautions to avoid fluorescent light baluns (the transformer in each light fixture), electrical wiring, heavy equipment (such as an elevator or other larger motor), and metal beams or metal walls. In general, the electrical wiring and metal items may be crossed (preferably at right angles), but not run alongside. Many authorities suggest that a 2-ft separation be maintained from these sources. EIA/TIA 569 provides specific guidelines.

Station wire must be of a category that conforms to applicable performance standards. Obviously, a Category 5 installation requires Category 5 certified cable. However, a Category 3 installation could actually use Category 4 or 5 certified cable, since either would exceed the requirement. In some cases, existing telephone station wire may be able to perform at the lower data rates of Category 3, which can be verified through testing. If the wire was installed after about 1984, it probably has enough twists per foot to approximate current Category 3 standards. Prior to that year, telephone twisted pair cable was sometimes manufactured with less than two twists per foot, unless it was made to meet more stringent AT&T premises distribution standards. A cable scanner that measures impedance and near-end crosstalk (NEXT) should give you an indication whether such cable will work adequately. You should be aware that none of the current EIA/TIA standards encourage you to "steal" pairs from an existing telephone station cable, but you may be able to retrofit an existing installation if absolutely necessary.

Station wire must also have an insulation type that meets safety and fire protection standards. Fire protection and safety standards are specified by Underwriters Laboratories (UL), the National Fire Protection Association's (NFPA's) National Electrical Code (NEC), Canadian Standards Association (CSA), International Standards Organization (ISO), and various European Community (EC) and individual country standards in other parts of the world. These standards are meant to exclude flammable cables (or ones that exhaust toxic gasses when placed in a flame) from any plenum spaces. A *plenum* is an air duct or a return air duct that is used for forced air circulation, such as from a heating or air conditioning (HVAC) system. Any air space used for forced air circulation, other than an actual room intended for work areas, is a plenum (which means that the area above drop-in ceiling tiles is plenum space if separate ducted air returns are not used). A common arrangement is to use a plastic return-air grid in place of a ceiling tile, or to use fluorescent lighting fixtures with a row of vent openings along the edges, to provide air return into the above-ceiling area and eventually back to the HVAC air handler. All of this space is air plenum and requires special plenum rated cables. Cables and cable ratings are covered in more detail in the chapters on Wiring Types and LAN Wiring Standards.

The wiring closet is the gathering point for all of the home-run cables that serve workstations of a particular floor or area in your building. The size and location of wiring closets are actually specified in standards, such as EIA/TIA 569. This standard refers to several types of wiring closets, such as the Telecommunications Closet (TC), the Intermediate Cross Connects (IC), and the Main Cross Connect (MC). Such a variety of wiring closets is appropriate for telephone wiring, since it is much more extensible than LAN cabling. However, since the topic of this book is restricted to LAN wiring, we will simplify this to just "wiring closet". For our purposes, then, all wiring outward from the patch or other station wire termination point to the workstation outlet jack will be considered simply horizontal cabling and will conform to those standard practices that are appropriate for LAN wiring.

As mentioned previously, horizontal cabling includes the wire terminations in the wiring closet. You will typically bring all the individual station cables into the wiring closet at the same entry point, then down a wall onto a wiring board. Wiring boards are often large 4-by-8 ft sheets of $\frac{3}{4}$-in plywood, mounted securely to a wall of the wiring closet, and painted an appropriate color. Of course, EIA/TIA 569 has a color code for every type of cross-connect, but you can probably use a neutral color of your choice on the wiring board. An alternate wiring closet arrangement uses free standing equipment racks or rails. Cables are routed in cable trays or in bundles across the top of the racks and down into each rack to the point of cable termination. This is particularly useful for very large installations where much equipment and many station terminations exist. Care should be taken in securing the cables and cable bundles to ceilings, walls, racks, and wire termination devices. In Category 5 installations, tightly binding the cables with tie-wraps should be avoided because it may deform the cable enough to distort its electrical characteristics. Likewise, you should maintain minimum bend radius with individual cables or with cable bundles. Avoid sharp corners and other abrasive edges.

Station wires may be terminated in the wiring closet on a punchdown block, such as a 66M or 110-type block, or they may be terminated directly onto a patch panel. If a punchdown block is used, cross-connect wire must be run between the station wire termination and any other connection device, such as a patch panel. One alternative is to use a punchdown block with a built-in connector (typically a so-called 50-pin Telco connector). A fan-out adapter cable, often called a *hydra* or *octopus cable*, connects from the connectorized punchdown block and splits to six (for standard 4-pair installations), eight, or even twelve 8-pin modular connectorized cables that can individually plug directly into a hub. To get from the punchdown to a patch panel, a 50-pin jumper cable could be used instead of the octopus cable. Some manufacturers also offer punchdown blocks with built-in 8-pin modular jack connections. Although some of these devices may be certified to Category 4 or 5, they are much more appropriate for Category 3 installations.

Direct patch panel termination is often used for Category 5 installations. This method provides the advantage of simple one-location termination. There are no cross-connects, punchdown blocks, jumper cables, fan-out adapters or Telco connectors. All of these are potential trouble spots for Category 5, since each wire termination introduces "untwist" and all components must be certified. Installation testing is also simplified with patch panel termination, because all station cables compose a Basic Link and can easily be scanned to certification standards. The only possible disadvantage to patch panel termination is that multiple-use application of the station cables is difficult, because each cable terminates into one patch panel location. Some manufacturers offer plug-in assemblies with a separate cable termination and plug-together jack arrangement, but this still limits the cable to two 8-pin modular four-wire jack terminations, at most. This disadvantage is probably not too serious, because it allows one to totally comply with the wiring standards, and special applications can always be handled as an exception with adapter cables.

Backbone wiring structure

All wiring between wiring closets is referred to as *backbone wiring*. In many cases, this backbone wiring will be between wiring closets on the same floor of a building, but the classic case is the multifloor model we saw in Fig. 1.1. Some standards refer to this as *vertical wiring*, because of the confusion with logical LAN backbone segments (which may or not run on backbone wiring). We will simply call it backbone wiring, and point out those few instances when we are referring specifically to a LAN standard's specification for logical backbone wiring.

Backbone wiring is usually simpler than horizontal wiring. It is customary to place LAN hubs or concentrators in each wiring closet to connect to all the workstation wires on served by that wiring closet. Backbone wiring then connects the hubs, one to another. Backbone wiring may be implemented in a daisy-chain (hub-to-hub-to-hub) fashion, or in a star fashion, as shown in Fig. 2.2. In a network topology that must guard against too many repeater jumps, such as Ethernet, the star connection for backbone wiring provides advantages. Since wiring closets in large buildings are frequently in a services shaft, one above the other, distances between wiring closets are minimal. Thus the backbone wiring from each floor may be concentrated in a single master wiring closet and the number of repeater jumps minimized. For other topologies, such as Token-Ring, it might be more convenient to daisy-chain the ring-in/ring-out from floor to floor.

In any event, the maximum cabling distances must still be observed. Backbone cabling may be terminated in the same manner as horizontal cabling. The backbone cable should be of a category and type that meets your

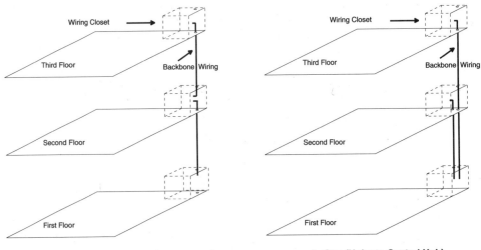

a. Daisy-chain (Hub to Hub)   b. Star (Hubs to Central Hub)

Figure 2.2 Backbone wiring may be implemented in a daisy-chain (hub-to-hub-to-hub) fashion, or in a star fashion.

performance requirements and the safety and fire protection requirements for your installation. If your installation requires Category 5 horizontal cable, for example, you should use the Category 5 cable for your backbone wiring. It does not hurt to use a higher category of cable than needed, so you might use Category 5 anyway. If you anticipate using your wiring system for multiple applications, double or triple each backbone wiring run. This will give you ample room for growth and the ability to rapidly respond to new requirements.

Some standards require that all riser cable (cable that goes vertically between floors) be riser rated, and some local authorities may require the tougher fire specifications of plenum cable. In any event, any openings you make between floors, whether in cable ports or direct opening, should have fire-stop material properly installed. This is particularly important in taller buildings. Consult the NEC, UL standards, and local standards for more guidance.

When wiring between far flung wiring closets or between floors, electrical grounding and bonding requirements should be observed. EIA/TIA 607 describes the proper practices in detail.

As fewer cables are used for backbone wiring, you may be tempted to terminate the cables directly into an 8-Pin modular male connector plug. Backbone cabling uses solid copper wire, and since most 8-pin modular plugs are designed for stranded wire, you could easily create a future problem. Just imagine an intermittent network failure that cures itself when you touch the backbone cable plug. Solid-wire 8-pin modular connectors are certainly available, but they are impossible to distinguish without close examination and most installers recommend against their use. A better approach is to simply use the same patch panel or punchdown that terminates the station cables to terminate the backbone cables. Mark the backbone cable patch position clearly to differentiate it.

You may wish to more clearly identify backbone wiring by using cable colors, cross-connect/patch panel colors, and special markings. These colors and markings, identified in EIA/TIA 569, are covered in later chapters.

If your backbone wiring actually involves floor-to-floor cabling in vertical shafts or cable ports, you may need to use special riser cable that has a fire-retardant sheath, certified to meet NEC low-flame requirements. The NEC requires that riser cable meet UL flammability tests to be used in high-rise buildings. Local regulations may supplement or amplify these requirements. Local building inspectors should be able to advise you about the requirements in your area.

As a last word on backbone wiring, fiber optic cable may provide some advantages over copper. Fiber optic cable is not subject to electrical or magnetic interference, and thus may be run in locations, such as elevator shafts or alongside power lines, that would cause problems for metallic cable. In addition, a cabling system may be electrically isolated between buildings or between floors of the same building using fiber optic cable. Metallic cable, on the other hand, must be carefully installed with adequate consideration of grounding and bonding. Fiber optic cable may also be used for multiple applications with the use of fiber optic multiplexers. Thus, it would be possible for one fiber optic connection to carry Ethernet, Token-Ring, T1, ATM, FDDI, and/or IBM-3270 all at the same time. Fiber optic cabling must meet the same plenum and riser specifications as metallic cable.

Location and Routing

The importance of location and routing of your cabling system components depends upon the level of cable facility that you intend to install. Category 3 installations have much more flexibility in routing than Category 5 installations. For networks that must provide 100 Mbps speeds, and have horizontal runs that approach the 90-m limit, you must be very attentive to installation details. Location and routing of the station cables are critical to high-performance operation.

Wiring closets should be located so that they are within 90 m "as the cable runs" from the proposed workstation outlets that are to be served. In some cases, the number of wiring closets may have to be increased or relocated to stay below this distance. A scaled plan drawing should be prepared and cable runs planned out and measured (on the drawing) to be certain. Allow an extra 3½ to 5 m for each vertical drop from the ceiling to the wall outlet and 2 to 3½ m for the drop from the ceiling to the patch panel. Remember that the vertical drop may start well above the ceiling grid, unless you choose to place the wire directly on the ceiling grid. If your patch panel is not wall-mounted, allow additional distance for the cable run to the rack or cabinet. Remember also that cable is normally run at right angles on the wiring board or in the rack. It is amazing how much extra cable all of these right angles require.

EIA/TIA 569 specifies the minimum size and suggested locations for "telecommunication spaces" such as wiring closets. It would be ideal if you could adhere to those standards, but some buildings may not allow such luxuries. Keep in mind that the network managers and technicians need to have quick and easy access to the patches and hubs in the wiring closets. If you absolutely must share space, place the wire termination point in a corner of the room that will not be blocked by storage or office equipment. Place it at an easily reachable level where there is adequate light, ventilation, and power. Several manufacturers make wall-mounted cabinets that conceal and protect your wiring and hubs.

All LAN cables should be routed away from potential sources of interference. Avoid fluorescent lights, motors, pipes, structural steel, and power lines. The National Electrical Code, TIA 568-A, and ancillary documents specify the separation distances, both for safety and for interference. With the exception of fluorescent lights and electric motors, most interfering conductors can be crossed at right angles, but you should still maintain a minimum separation for safety reasons.

Some authorities advocate direct physical support of Category 5 LAN cables.

This would require the use of cable trays, ducts, or conduit for the entire route of each cable. In many situations, these extensive precautions are simply impossible or cost-prohibitive. However, in large buildings with suspended ceiling grids, it may be desirable to use a cable tray routing system to help organize and protect the LAN cable. The key element is to eliminate any deformation in the insulation or twist integrity of the cable. That generally means that you must avoid long unsupported spans of wire, sharp bends, kinks, binds, nicks, and tight tie-wraps.

A Simple Design

To illustrate a basic data-only wiring system design, we will show a cabling plan for a simple two-story building using the concepts we have covered in this chapter.

Our example will have 24 workstations per floor. We will use Category 5 specifications, direct patch panel termination, and will have one wiring closet per floor. We will specify the components, installation, and testing and will develop a sample bill of materials.

Figure 2.3 shows the layout of each floor. We will specify flush-mounted jacks for the user outlets. Each user will have a 3-m Category 5 cable from the wall

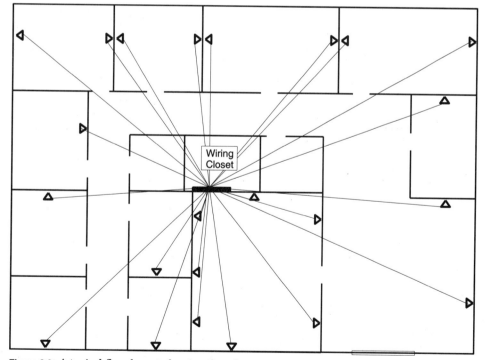

Figure 2.3 A typical floor layout showing the telecommunications outlets and direct cable runs to the wiring closet. You should allow 25-50% more cable for cable trays and obstacle avoidance.

TABLE 2.1 Typical Bill of Materials

Quantity	Description
A/R	Cable, 4 twisted pair, 24 gauge, Category 5, plenum-rated
A/R	Cable, 4 twisted pair, 24 gauge, Category 5, riser-rated, gray or white
48	Plate for modular jack, ivory
48	8-pin jack module, Category 5, wired per T568A, ivory
2	Patch panel, 24 position, Category 5
2	Mount for patch panel
48	User cord, 3 m, Category 5, stranded twisted pair, PVC (non-plenum), gray
48	Patch cord, 1 m, Category 5, stranded twisted pair, PVC (non-plenum), yellow
2	Duct, fingered, 3 in × 3 in, 6-ft length (cut to 2-ft lengths for patch cord management)
A/R	Duct, fingered, 2 in × 2 in, 6-ft length (for vertical cable entrance into wiring closet/patch panel area)

NOTE: Quantities shown are for two floors, 24 work areas per floor, and one LAN cable per work area. Some standards, such as TIA-568A, recommend two cables per work area. An existing telephone cable at each work area would yield the total of two cables.

outlet to the workstation. We will use Category 5 plenum cable with 4 24-gauge twisted pairs for the station cable that runs from each user outlet to the wiring closet. At the wiring closet, we will use a wall-mounted 24-position patch panel that will directly terminate each station cable. A 1-m Category 5 patch cord will be provided to connect each patch position to the network hub or MAU. In some situations, longer patch cords should be specified. Plan the placement of your hubs before ordering patch cords. Between the two floors we will run one riser cable of the same type as the station cables, if local codes permit.

Of course, the LAN network will also need appropriate LAN hubs or MSAUs, network adapters, servers, routers and software, but our concern here is simply the wiring system that connects all these devices.

The bill of materials is shown in Table 2.1. If you have to obtain the materials, you will also have some brand-name decisions to make. Some brands have special features, offer long warranties, or are easier to install. An experienced installer will have preferences that can guide your choices. There will also be some differences in quality and pricing. All the items should be certified by their manufacturers to meet the Category 5 performance criteria you have set. To implement the two-cables-per-workstation suggestion of TIA 568-A, you will need to double many of the quantities.

The wall plates, user and patch cords, and cable are generally available in a variety of colors. The wall plates and user cords are usually color coordinated to the office. The color choices are somewhat limited, so you may have to go with a neutral color such as gray or ivory. Surface raceway is also available in color choices. Patch cords technically fall under the miscellaneous category of EIA/TIA 569, and are ideally yellow in color. However, if your wiring closet is actually a vacant wall in a common area, and a sea of yellow wire would distract from the decor, you may wish to specify a more neutral color. The EIA/TIA police will never know (we hope).

The instructions to the installers will be as follows. Pull the station cables as home runs, with no splices, from each workstation outlet location to the wiring closet on each floor, avoiding electrical wires, pipes, motors, and florescent light fixtures. The station cables will be run from each outlet through hollow walls to the space above the ceiling grid. Outlets may be surface mounted on solid walls and the cable placed in surface conduit to reach the area above the ceiling grid. Cables should be run supported as much as possible with no tight tie-wraps, no kinks, and respecting the minimum bend radius for Category 5 (approximately four times the cable diameter). Cable runs must not exceed 90 m from outlet to patch. The patches should be mounted at a convenient height on the wall of each wiring closet with adequate support for the cable bundle and no tight tie-wraps. The riser cable should be run in an existing cable port in a vertical shaft or utility area, if possible. Fire-stopping material should be placed where the riser cable runs between floors.

The cable pairs should be wired using the T568A pattern of TIA 568-A. Proper workmanship standards should be observed, including the minimum twist standard for connecting Category 5 cable. After installation, each station cable connection should be scanned from outlet to patch with a Category 5 cable scanner that complies with TSB 67. If a cable run is near the 90-m maximum length, the entire Channel should be scanned, including the patch and user cords that are to be used for that workstation connection. A printed report will be provided, including a plan diagram showing the numbering of all outlets. All cables, outlets, and patch panels are to be marked in accordance with EIA/TIA 606.

Well, that pretty much covers an ideal, concise wiring system specification. You could mention a few more standards, if you wish. As you will see in later chapters, there are quite a number of standards that more closely define your installation. Keep in mind that most of these standards are voluntary and exist to make absolutely certain your wiring system will be up to stringent performance specifications. Many of the standards overlap or are "coordinated," so that they actually say the same thing. Too much reference to standards may actually raise the price of your installation. Some of the standards are overkill for small simple networks. For example, it is doubtful that you could make use of half the detailed marking specifications in a 10- or 20-user network. However, if you are planning a 1000- or 2000-user wiring system, adherence to the marking standard will be extremely useful. Likewise, the color codes and even the routing guidelines may be less significant for smaller networks with shorter runs. You will have to be the judge. The standards do not imply that your network will fail if you do not observe all the guidelines; they simply say that it will work at the specified performance category if you do.

Chapter

3

Wiring Types

Many types of wire and cable are used in LAN wiring. This chapter will describe the major types of cable that are currently in use. You should define the cable that will be used long before you begin your cable design. The cable chosen will often dictate the connectors, terminations, useful distances, routing and LAN types that your wiring will support.

We will cover types of cable that are used for all types of networks, including Ethernet, Token-Ring, and Arcnet. Although we mention these network topologies and their variations, we will concentrate here on the cable, rather than on its use.

Basic Cable Types

All LAN cable can be divided into three basic types: twisted pair, coax, and fiber optic. We will cover each of these types in detail. We will also cover round and flat nonpaired wire often used for telephone sets. However, because nonpaired wire is not used for LAN wiring, it will be presented only so that you may easily recognize it and avoid its use.

Each basic cable type has several variations and different types may even be found in combination. An example of this is twisted pair cable, which may be found in unshielded and shielded varieties. There are also several types of nonpaired wire, such as round 4-wire cable, ribbon cable, and flat cable (often called "silver satin"). Some manufacturers, such as IBM, specify combination cables that contain both shielded and unshielded pairs. Even combination fiber optic and metallic cables are offered. However, for our purposes here, we will avoid all of the myriad of combinations and concentrate on the pure cable types.

UTP and STP Cable

Twisted pair wiring is the most common type of LAN wiring cable in use today. It is versatile, easy to install, inexpensive, and has favorable performance characteristics. Twisted pair cable is available with or without shielding. It comes in a variety of colors, wire gauges, insulation, twisting, and outer sheath materials.

An illustration of twisted pair wire is shown in Fig. 3.1. A large number of pairs can be in the same cable sheath (the outer jacket). Telephone cables are commonly available in 2 pair, 4 pair, 6 pair, 25 pair, 100 pair, and even larger bundles. LAN twisted pair cable, however, usually comes as 4 pair cable.

The 4 pair cable is a topic of several standards, including TIA 568-A. It is the cable around which most of the important LAN cable specifications and performance tests are based. Most LAN topologies actually use only two of the four pairs, however, so some installations place two LAN connections on each 4 pair cable. Telephone connections often use two pairs and may sometimes be wired in the same cable with a 2 pair LAN connection. You should be cautious in robbing pairs from the LAN cable, because some of the 100 Mbps and higher LAN schemes may use all four pairs.

Another common cable that is found in twisted pair wire installations is the 25 pair jumper cable. This cable is preterminated in 50-pin male or female connectors (sometimes called a 50-pin Telco connector). In telephone wiring, the 25 pair jumper cable is convenient for connecting between the local exchange carrier's RJ-21X demarcation point and the user's punchdown blocks or Private Branch Exchange (PBX) switching equipment. In LAN wiring, the 25 pair jumper is commonly used between connectorized punchdown blocks and patch panels. Category 3 and Category 5 grade 25 pair jumpers are available, but some caution should be used in deploying them in a Category 5 installation. Some authorities are concerned with the combination of two or more 100 Mbps signals in the same cable sheath. A 25 pair jumper would theoretically allow you to combine as many as six such signals within the same sheath (6 × 4 pairs = 24 pairs with one unused pair). Also, each connector introduces more untwisted wire into a circuit and some of the standards limit the amount of untwist as well as the number of connectors in a link.

General construction

Twisted pair cable consists of one or more pairs of insulated wires that are twisted together and joined in a common sheath. The main characteristics of twisted pair cable are wire gauge, stranding, twist pitch, insulation type, characteristic impedance, and sheath material. Each of these items may affect the suitability of a cable for a particular application.

Cables for LAN wiring may be made with either stranded or solid copper wire. Solid wire is normally used for cable runs that will be terminated on

Figure 3.1 Illustration of twisted pair wire.

insulation-displacement connectors (IDCs), such as outlet jacks or punchdown blocks. Stranded wire is normally used for user and patch cords that are terminated with an insulation-piercing 6-pin modular (RJ-11 style) or 8-pin modular (RJ-45 style) modular connector (plug). You should never use solid copper wire with a modular-type plug, unless the plug is specially designed for solid wire (see Chapter 10). Some older installations for Category 3 and below use screw terminal jacks that technically can be used for stranded wire, although solid wire is the norm for station cable.

The twisting of the two wires causes interfering electromagnetic fields to couple equally to each wire in the pair, as shown in Fig. 3.2, thus effectively canceling out the resulting interfering signal. This mode of operation is referred to as a "balanced" transmission line. For proper cancellation it requires that a desired signal be applied to the wire pair by a balanced driving circuit and load. The equally-coupled interfering signal is ignored by the balanced load, although a common-mode component of the interfering signal may exist. A balanced circuit is shown in Fig. 3.3. As an additional benefit, electromagnetic emissions from a balanced twisted pair are reduced (note we said "balanced!"). This prevents the high frequencies of a LAN signal from interfering with other devices. However, some radiation of the signal does occur. The transmitted signal amplitude is kept low to maintain the spurious emissions within acceptable limits. Longitudinal balance of the entire Channel is important for low emissions as well as susceptibility to outside interference.

Wire sizes

The conductor diameter in twisted pair wires is referred to by wire gauge. The common standard is the American Wire Gage (AWG). Table 3.1 shows a comparison of wire gauges to physical size. Smaller diameter conductors correspond to larger wire gauge numbers. Thus, AWG-26 is smaller than AWG-19. Telephone outside plant cabling is typically 24 or 26 gauge, although in rural areas 19 or 22 gauge may be used. The larger diameter wires have more physical strength, which is an advantage for wires that must be run between distant supports. Larger wires also have a lower resistance per unit length, which reduces resistance to direct current and lessens attenuation to alternating frequencies, such as voice. (At LAN frequencies, the capacitance is the primary contributor to attenuation.) The greater copper content of larger wire sizes also increases the cost of the wire.

Figure 3.2 Twisted wire causes interfering electromagnetic fields to couple equally to each wire in the pair and cancel out at the load.

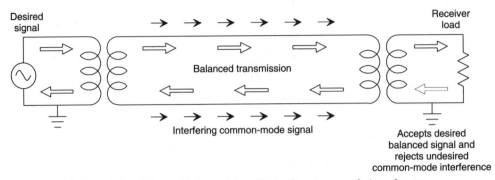

Figure 3.3 A balanced circuit is used to cancel the effects of common-mode interference.

At higher frequencies, such as those of a high-speed LAN, signal current concentrates at the outer diameter of the conductor, a phenomenon known as *skin effect*. In a stranded conductor, the outer skin is ill-defined, because of the many conducting strands that make up the wire, and attenuation is increased. For that reason, standards typically limit the length of stranded wire that is allowed in a Channel. Solid copper wire must be used for the horizontal cable portion of the Channel. Stranded wire may be used for user and patch cords, where a 20% increase in attenuation is allowed.

The common wire gauge that is specified by standards for LAN wiring use is 24 gauge solid copper wire. This wire size provides a good compromise between attenuation and cost. It is also an easy size to work with.

Electrical characteristics

Twisted pair wire, as with other types of transmission lines, has a characteristic impedance. This impedance is a result of the dielectric properties of the insulation and the closeness of the conductors. The insulation's dielectric properties are a function of the type of material used. A variety of characteristic impedance values are available. The wire used for LAN twisted pair wiring has a characteristic impedance of 100 ohms, 120 ohms, or 150 ohms. Most of the UTP cable you will deal with will be rated at 100 ohms, as that is the standard for unshielded twisted pair wire in most parts of the world.

Shielded, twisted pair (STP) wire generally has a characteristic impedance of 150 ohms. STP is often found in Token-Ring installations that use the IBM cabling system. To connect between 100 ohm and 150 ohm wire, you must use a balun to minimize reflections caused by the impedance mismatch. If network interface components for 150 ohm STP are used, an adapter cable with a built-in balun must be used to couple to 100 ohm UTP station cable. This arrangement is shown in the chapter on cabling systems. The specific type of STP cable that is defined in TIA 568-A is called *STP-A* to differentiate it from other types of STP cable.

If you are familiar with telephone circuits, you may know that such circuits are generally characterized for operation at 600 or 900 ohms. This load impedance is designed to minimize current flow to subscriber telephone instruments, in addition to other factors. It is not representative of the characteristic impedance of the wire pair, which is not as significant at voice frequencies.

Why is the characteristic impedance important to LAN wiring? The reason is twofold. First, the maximum amount of power is transferred when the impedances of the source and load are matched. A greater power transfer extends the usable cable distance. Second, mismatches cause reflections of the signal, which can deteriorate the quality of the received LAN signal and cause unwanted emissions. At high LAN data rates and long cable distances, this might become critically important to performance.

As we will see later, a cable's characteristic impedance may vary over the frequency range of the LAN signal. The standards specify a maximum variation that must be met for the cable to meet minimum performance criteria.

The resistance and capacitance of a twisted pair cable cause an attenuation of the LAN signal that is proportional to length of cable. This signal attenuation is expressed in decibels (dB), indicates a power ratio, $10_{log}(P_{in}/P_{out})$, of the input vs output power expressed as a logarithm. The typical attenuation of a cable is given in dB per foot or meter so that you can calculate the total attenuation of a cable run by multiplying its per unit value by the total length. Fortunately, a standards-compliant installation will insure that the total attenuation (including cable and connectors) is within the maximum limits necessary for proper LAN operation. The attenuation of a typical cable run is enough below the maximum allowed that the an attenuation measurement usually reveals bad connections or partial cable breaks, rather than cable deficiencies.

TABLE 3.1 American Wire Gage for Selected Solid Bare Copper Wire Sizes

Gage (AWG)	Nominal diameter (mm)	Nominal diameter (inches)	Area (circular mils)	Nominal resistance (ohms/kft)
10	2.60	0.1019	10380.0	0.9989
12	2.05	0.0808	6530.0	1.588
14	1.63	0.0641	4107.0	2.525
16	1.29	0.0508	2583.0	4.016
18	1.02	0.0403	1620.0	6.385
20	0.813	0.0320	1020.0	10.15
22	0.643	0.0253	640.4	16.14
24	0.511	0.0201	404.0	25.67
26	0.404	0.0159	253.0	40.81
28	0.320	0.0126	159.8	64.9
30	0.254	0.0100	100.5	103.2
32	0.203	0.0080	63.21	164.1
34	0.160	0.0063	39.75	260.9
36	0.127	0.0050	25.00	414.8

SOURCE: National Bureau of Standards

The capacitance of twisted pair wiring is the factor that causes attenuation to increase steadily with frequency. The attenuation caused by capacitance is proportional to $1/(2\pi f)$. Thus, the greater the frequency, the less the parallel impedance between the wires, and the more the signal attenuation. Although this attenuation factor is linear, in practice, twisted pair cables exhibit peaks and valleys that can affect transmission performance at specific frequencies. For this reason, the new performance measurement standards for installed cable require numerous discrete measurements over the entire frequency range of a cable category.

Installation practices can adversely impact twisted wire cable performance. The impedance characteristics can be significantly disturbed by any distortion of the cable insulation or cable twist. These disturbances can be easily observed on a device called a time-domain reflectometer (TDR). When connected to a TDR, a normally flat response of a cable will get all sorts of lumps and bumps when the cable is sharply bent, stapled, tie-wrapped, and partially untwisted. These distortions in the cable may cause the installed cable to fail the final installed testing and should therefore be avoided.

Insulation

Two primary types of insulation materials are traditionally used in twisted pair LAN wiring, polyvinyl chloride (PVC) and fluorocarbon polymers. PVC is very flexible and is often used with stranded wire for LAN user cords and patch cords, where its flammability is not a disadvantage. PVC has dielectric properties that may make it unsuitable for wire insulation of Category 5 cable, although it can be used in cable jackets. Polyethylene (PE) is sometimes used for Category 5 cable and may be substituted on one pair of some plenum-rated cables. Fluorocarbon polymers include polytetrafluoroethylene (PTFE or TFE) and fluorinated ethylene-propylene (FEP), which are not as flexible as PVC, but meet the stringent flammability testing for use in plenum spaces. These fluorocarbon polymers are sometimes referred to by the DuPont trademark Teflon®. Wire insulation and sheaths may be a copolymer of TFE and FEP. Another fluorocarbon polymer that is often used in cable jackets of plenum-rated cables is ethylene-chlorotrifluoroethylene (ECTFE or HALAR®). This material does not have the proper dielectric qualities to be used as a wire insulation, but efforts are underway to formulate it for that purpose.

The increasing use of large amounts of cable for LAN wiring has resulted in the creation of new cable guidelines that emphasize the flammability of the cable sheath and insulation. Conventional wire insulations were found to be relatively flammable and a hazard when placed in the air plenums. Plenums, or air ducts, carry the heated or cooled air around buildings. Any cable that burned in a plenum could create toxic gas and smoke that would be a danger to people in other parts of a building. In addition, the fire could actually spread through plenums. Manufacturers soon developed cable insulations that were less flammable and could be used in plenums.

Early fire protection specifications referred to sections of the National Electrical Code (the NEC) or to flammability tests of Underwriters Laboratories (UL). Similar specifications were published by other countries. Several of the pertinent NEC articles are shown in Table 3.2. The NEC Article 800 is often referred to for LAN cabling and telecommunications cable. The NEC differentiates cable types by voltage class and by flammability. Class 2 (up to 150 volts) and Class 3 (up to 300 volts) are both covered by Article 725. Article 725 might appear to apply to LANs, but it actually refers the user to Article 800 for cable classified for communications use (which does include LAN cable). In addition, permitted use is graded by three levels of flammability: general use, riser use, and plenum use.

The confusing labeling of NEC and UL specifications led to a revision of the NEC in 1987 that instituted a system of alphanumeric designators for permissible cable-use ratings. The common cable-use codes are shown in Table 3.3. Cables may be specified that meet or exceed the requirements for a specified use. You may find cable that has dual markings, if the cable is certified for use in each classification.

The FEP Crisis

The impact of computer networking on the plastic materials industry has been felt very directly by users of plenum-rated cable. Plenum cable uses insulation and sometimes jacketing made with fluorocarbon plastics for their fire-retardant properties. Fluorinated ethylene-propylene (FEP) is often the material most used for insulating the individual conductors because of its superior dielectric properties at the very high frequencies of modern LAN cable.

The problem is that the increasing use of this popular plenum-rated material has outpaced the ability of the material manufacturers to produce sufficient quantities for the cable industry. Additionally, the rapid growth of LANs and the tendency to over-wire new installations for future requirements has thus far made increases in manufacturing capacity only a short-term cure. Of course, some double-ordering from cable and material users has occurred, which has caused the appearance of shortages to be even more severe.

For the cable user, the FEP crisis may mean that supplies of cable are lower than normal, or that the favored manufacturer, style, or even jacket color is not available on short notice. Deliveries may be extended and selection limited.

Several cable manufacturers have substituted other types of insulation on one of the four pairs in popular LAN cables to stretch the supply. Unfortunately, they are very restricted in the types of material that can be used for the very high speed data rates of modern LAN systems. Common substitutions are polyethylene, polyolefin, and some special polyvinyl chloride compounds. Some of these materials, in pure form, would not be able to pass the UL flame-spread tests, but are acceptable when used in conjunction with three normal FEP pairs and a plenum-rated jacket.

Costs of plenum-rated cables of all types have increased because of the material shortage. Some question whether the substitution of alternate materials may affect the long-term stability of the cable. However, there is no evidence of any problems at this point, as the alternative polyethylene and polyolefin materials are normally quite suitable for high frequency use in nonplenum applications. Experimentation is ongoing to find other materials, such as from the chlorofluorocarbon family, that can meet plenum requirements and have adequate dielectric characteristics for Category 5 LAN use. You should start to see some of these materials hit the market over the next several years.

At the current rate of growth of LAN networks, we may experience periodic material shortages for several years to come.

TABLE 3.2 National Electrical Code (NEC) Articles for Low-Voltage Wiring

Article	Types of cable
725	Remote signalling and power limited circuits.
760	Fire protection signalling systems.
770	Fiber-optic cables.
800	Communication cables.
820	Coax cables.

Color-coding and marking

Each insulated wire of a twisted pair LAN cable is colored differently. The colors form a standard code, so that each wire may be easily found and terminated properly. The color code for 4 pair LAN cables is shown in Table 3.4. Each pair of the cable has complementary colors. For example, pair 1 wires are coded white-blue and blue-white. The white-blue wire is a white wire with a blue stripe at intervals along the insulation of the wire. (The stripe is sometimes called a *tracer*.) Conversely, the blue-white wire is a blue wire with a white stripe. In a 4 pair cable, white will be the common color of all the wires, and the white wire will always be numbered (or punched down) first. Each pair of a 4 pair wire may be referred to by its color that is unique from the other three pairs. Thus, the "blue" pair contains the white-blue and the blue-white wires, while the "green" pair contains the white-green and the green-white wires.

Color-coding and proper termination of each wire color is very important in LAN wiring because the signals are polarity sensitive; reversing a pair will cause a failure. The polarity of each pair is often referred to as *tip* and *ring*, which stems from the days of telephone plug-boards used to route calls. The switchboard plug consisted of three contact areas, referred to as the tip, ring, and sleeve, much like a modern stereo plug. The primary color was wired to the ring and the secondary to the tip. The sleeve was used for grounding.

The 25 pair jumper cable is another common cable that may be found in some LAN installations. The color code for 25 pair cables is shown in Table 3.5. Although 25 pair and larger bulk pair cabling is often used in telephone wiring, ordinary telephone cabling is not appropriate for today's' very high-speed data rates. It can function at Category 3 as a multi pair jumper at short distances, such as between a patch panel and a connectorized punchdown block (both with 50-pin telco jacks). If 25 pair cable is used in a LAN application, it should have the proper twists-per-foot as cable for Category 3. Some manufacturers offer 25 pair jumpers that are said to meet Category 3 or Category 5, but you should use caution if your ultimate application will need to support 100 Mbps or higher data rates. Placing multiple 100 Mbps circuits within the same 25 pair sheath may cause excessive crosstalk, resulting in excessive error rates or even link failure. The TIA is working on specifications for 25 pair cable which will allow its use for multioutlet centers, where modular furniture is used.

TABLE 3.3 Common National Electrical Code (NEC) Cable-use Codes for Metallic Communications and Control Cable

NEC article	Code	Meaning	Allowable substitutions*
725	CL3P	Class 3 Plenum	MPP CMP FPLP
	CL3R	Class 3 Riser	CL3P MPR CMR FPLR
	CL3	Class 3	CL3P CL3R MP MPG CM CMG FPL PLTC
	CL3X	Class 3, limited use	CL3P CL3R CL3 MP MPG CM CMG FPL PLTC CMX
	CL2P	Class 2 Plenum	CL3P
	CL2R	Class 2 Riser	CL3P CL2P CL3R
	CL2	Class 2	CL3P CL3R CL2P CL2R CL3 MP MPG CM CMG FPL PLTC
	CL2X	Class 2, limited use	CL3P CL3R CL2P CL2R CL2 CL3 CL3X MP MPG CM CMG FPL PLTC CMX
800	MPP	Multipurpose Plenum	None
	MPR	Multipurpose Riser	MPP
	MP, MPG	Multipurpose	MPP MPR
	CMP	Communications Plenum	MPP
	CMR	Communications Riser	MPP CMP MPR
	CM, CMG	Communications	MPP CMR MPG MP
	CMX	Communications limited use	CMG CM

*Note: In general, a cable with a more strict usage code may be substituted in an application that allows a less strict usage code. For example, a cable rated to the stricter Class 2, Plenum (CL2P) code may be substituted in Class 2, Riser (CL2R) application. Likewise, CMP, MPP, and FPLP, which may be substituted for CL3P, will substitute for any code that allows CL3P to be substituted. The chart does not list all possible substitutions, only those to the next level.

SOURCE: 1996 NEC

Cable sizes above 25 pairs are usually in multiples of the basic 25 pair bundle. Each group of 25 pairs is called a *binder group* and is marked within the larger bundle with a uniquely colored plastic binder that is spiral-wrapped along the bundle. Groups of four or more binder groups may be wrapped and included in

TABLE 3.4 Color Code for 4-Pair Cable Pairs

Pair #	Primary color ring	Secondary color (stripe) tip
1	blue	white
2	orange	"
3	green	"
4	brown	"

TABLE 3.5 Color Code for 25-Pair Cable Pairs

Pair #	Primary color *ring*	Secondary color (stripe) *tip*
1	blue	white
2	orange	"
3	green	"
4	brown	"
5	slate	"
6	blue	red
7	orange	"
8	green	"
9	brown	"
10	slate	"
11	blue	black
12	orange	"
13	green	"
14	brown	"
15	slate	"
16	blue	yellow
17	orange	"
18	green	"
19	brown	"
20	slate	"
21	blue	violet
22	orange	"
23	green	"
24	brown	"
25	slate	"

the same outer jacket to form larger cables. The sheath of twisted pair cabling should be clearly marked by the manufacturer to show the level of performance that is guaranteed. The marking normally shows three performance indicators, the EIA/TIA Category, the UL/NEC permitted use, and the number and size of conductors. The manufacturer's name and part number are also normally shown. A typical marking might be:

4PR, 24GA, CMP, Cert. to EIA / TIA Category 5, XYZ CABLE CO., PN1999

This example would be a 4 pair 24 AWG cable, tested for fire and voltage rating to CMP standards, and certified to EIA/TIA Category 5 performance standards.

Because standards for wire and cable have changed so much over the past several years, there might be several variations that actually mean the same thing. For example, the marking might say "Meets UL 910" or "Article 800" for fire-protection designation, rather than the CMP wire use type. Similarly, because the TIA 568-A standard has been in a state of refinement, you might see markings such as "TSB-40" or the older "Level 5." If the cable is marked simply "EIA/TIA 568" or just "Category 5," rather than "TSB-40" or "TIA 568-A," you may not be getting cable that meets the most current standards, which may be more rigorous in some cases.

Shielding

Shielding is sometimes used with twisted pair cable to provide better noise immunity and lower emissions. Two types of shielding are commonly available for STP cable: the foil shield and the braided wire shield. The foil shield usually has a bare stranded wire, called a *drain wire*, that is in contact with the shield and provides the electrical connection for the foil. Braided shields may have a drain wire, or the braid may be formed into a wire or clamped to make electrical contact. Some cables may have combination foil/braid shields or double shields.

Coax Cable

Coaxial cable, or coax, is the original LAN cable. It was first used for local area networking in Ethernet networks, IBM PC-NET broadband networks, and Arcnet networks. Coax is still in use in thousands of locations, even though many newer installations have converted to twisted pair.

Coax is primarily used for its self-shielding properties, low attenuation at LAN frequencies, and installation expense. The construction of the cable greatly reduces susceptibility to outside interfering signals and noise, as well as minimizing the radiated emissions from the cable. Because coax was originally designed to carry radio frequencies, it has fairly low attenuation characteristics. The cable is readily available, relatively inexpensive, and allows a daisy-chain or tapped LAN connection that minimizes total cable length.

Coax cables may have solid or stranded center conductors, foil or braided shields, several types of insulating dielectric, and several types of outer jacket materials. The cable is available in a variety of standard sizes and impedances.

Several types of coax cable are available for LAN use. Table 3.6 shows the common types and their uses. As with twisted pair cable, coax is available in a variety of insulation types and may be used in plenum spaces and riser shafts. Coax is included in IEEE 802.3, Ethernet Version 2.0, the NEC, and other standard documents. It was formerly included in EIA/TIA 568 1991, but was dropped in TIA/EIA 568-A 1995 as a recommended cable type.

General construction

Coax cable gets its name from its construction. As you can see from Fig. 3.4, coax consists of a center conductor and a coaxially positioned outer shield conductor that are separated by an insulating plastic, called a dielectric. An outer jacket insulates the shield. The shield may be a foil-wrap with a drain wire or a wire braid. Some coax, such as that used for thick Ethernet, may have a double shield layer.

Theoretically, the positioning of the center conductor, surrounded completely by the concentric shield conductor, keeps all electromagnetic fields between the two conductors, as shown in Fig. 3.5. This mode of operation is referred to as "unbalanced", as opposed to the balanced arrangement of twisted pair cable. Other terms for the unbalanced mode include single-ended and bipolar.

TABLE 3.6 Common Types of Coaxial Cable

Coaxial cable type	Rated use	Typical NEC type	Center cond. AWG	Nom. imped. (ohms)	Nom. vel. prop. (c)	Typical O.D.	
						mm	in
RG-6/U	Nonplenum	CL2	18 sol.	75	82%	6.86	0.270
	Plenum	CMP	18 sol.	75	82%	5.94	0.234
RG-8/U	Nonplenum	CL2	11 str.	50	78%	10.24	0.403
	Plenum	CMP	10 str.	50	83%	9.07	0.357
RG-11/U	Nonplenum	CM	14 sol.	75	78%	10.29	0.405
			18 str.	75	66%	10.29	0.405
	Plenum	CMP	14 sol.	75	83%	8.84	0.348
RG-58/U	Nonplenum	CL2X	20 sol.	50	66%	4.90	0.193
	Plenum	CMP	20 sol.	53.5	69.5%	4.04	0.159
RG-58A/U	Nonplenum	CL2	20 str.	50	66%	4.90	0.193
	Plenum	CL2P	20 str.	50	80%	4.06	0.160
RG-59/U	Nonplenum	CM	20 sol.	75	78%	6.15	0.242
	Plenum	CMP	20 sol.	75	82%	5.38	0.212
RG-62/U	Nonplenum	CM, CL2	22 sol.	93	84%	6.15	0.240
	Plenum	CMP	22 sol.	93	85%	5.08	0.200
Ethernet 10Base2	Nonplenum	CL2	20 str.	50	80%	4.70	0.185
	Plenum	CL2P	20 str.	50	80%	4.06	0.160
Ethernet 10Base5	Nonplenum	CM, CL2	12 sol.	50	78%	10.29	0.405
	Plenum	CMP	12 sol.	50	78%	9.53	0.375

Note: In general, a cable with a more strict usage code may be substituted in an application that allows a less strict usage code. For example, a cable rated to the stricter Class 2, Plenum (CL2P) code may be substituted in a Class 2, Riser (CL2R) application. Likewise, CMP, MPP, and FPLP, which may be substituted for CL3P will substitute for any code that allows CL3P to be substituted. The chart does not list all possible substitutions, only those to the next level.

The shield is maintained at "ground" potential, while the center conductor is driven with the LAN signal.

Because the shield is grounded, interfering signals from outside the coax cable should be prevented from entering the cable and coupling to the center conductor. Grounding is quite important in coax cable installations. The shielding properties of the cable depend somewhat upon good ground connections (both signal and earth grounds). However, the interconnection of grounds from different parts of a building, or even between buildings, can cause problems. Grounding requirements for thick Ethernet coax are fairly severe, to ensure proper operation. Safety and signal integrity are also issues in grounding. The NEC and EIA/TIA 607 detail grounding and bonding requirements for coax LAN cable.

Ethernet LAN wiring uses two types of coax cable, often referred to as *thicknet* and *thinnet*. Thicknet is the original Ethernet coax trunk distribution cable, now called 10Base5. Its larger diameter gave rise to the nickname thicknet after much smaller coax began to be used. Thicknet cables are run in the walls or above the ceiling tiles near locations where workstations are planned.

A transceiver mounts directly around the coax and makes connection to the center conductor through a hole drilled through the outer jacket, shield, and dielectric insulation. This arrangement is sometimes referred to as a *vampire tap*.

Thinnet is a newer Ethernet coax cable standard, 10Base2, that allows connection to the back of the workstation adapter. The adapter, in effect, contains a built-in transceiver. Thinnet uses BNC-type connectors and T-adapters. The cable is about ¼ inch (1 cm) in diameter and is much more flexible than thicknet cable. It is also less expensive, giving rise to another nickname, *cheapernet*. Distances for a thinnet Ethernet segment are limited, as compared to thicknet.

Arcnet is another LAN topology that originally used coax cable. Several types of coax have been used with Arcnet, including RG-11/U, RG-59/U, and RG-62/U. RG-11/U and RG-59/U are 75-ohm coax cables, while RG-62/U is a 93-ohm cable. The workstations are connected directly to the coax in a star arrangement. Each leg terminates in a passive or an active hub device that couples the LAN signal to all of the other connected workstations or servers.

Wire sizes

Wire sizes for coax cable are usually specified by "RG" number or manufacturer's part number. Thick Ethernet cable is a special case that is referred to by the names Ethernet trunk cable, backbone cable, or 10Base5 cable. A list of coax types used in LAN wiring is shown in Table 3.7. The RG numbering system is a standard coax cable rating system that identifies physical size, characteristic impedance, power handling, and other characteristics. Common LAN coax cables are Ethernet, RG-8/U, RG-11/U, RG-58A/U, RG-59/U, and RG-62/U. The outer jacket of RG-8/U and RG-11/U is roughly ½ inch (12.7 mm), while RG-58A/U, 59/U and 62/U are on the order of ¼ inch (6.4 mm). Plenum versions of these cables are usually a little smaller. The wire sizes of the center conductor and the diameter of the outer jacket (including dielectric and shield) are defined by the RG number and the insulation type. You must specify whether you want connectors for PVC or plenum insulation types, since the cable dimensions are not the same.

Figure 3.4 Coax consists of a center conductor and a coaxially positioned outer shield conductor separated by an insulating material, called a *dielectric*.

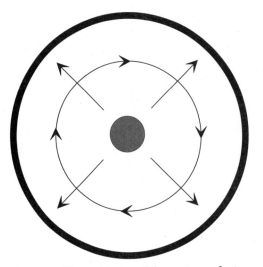

Figure 3.5   The positioning of the center conductor, surrounded completely by the concentric shield conductor, keeps all electromagnetic fields between the two conductors.

Electrical characteristics

Coax cables are available in a variety of standard characteristic impedances. Because the primary use for coax is the transmission of RF signals, the coax impedances often reflect the needs of RF equipment. These standard impedances are 50 ohms, 75 ohms, and 92 ohms. The dimensions and properties of the center conductor, dielectric, and shield combine to define the coax cable's characteristic impedance. The dielectric constant in flame-resistant cable is very different from that of conventional polyethylene or foam dielectric cable. It is for this reason that plenum cable typically has a smaller cable diameter.

The attenuation of coax cable less than 1.5 dB per 100 ft at 10 MHz. At 100 MHz, the typical attenuation is under 5 dB per 100 ft. Obviously, the useful coax cable length is reduced as the frequency goes up. Some types of networks base their maximum coax cable runs on the total allowable loss from end to end (the loss budget), while other types of networks have timing constraints that limits the usable distance.

As with other wiring types, the capacitance of the cable is the primary contributor to attenuation at higher frequencies. Even so, coax has a greater maximum usable length than twisted pair. For example, a 10 Mbps Ethernet segment can be 500m (1640.5 ft) using thicknet and 185 m (607 ft) using thinnet. (Note: segment lengths in Ethernet are limited by attenuation and frequency-based signal distortion, while network span is limited by timing constraints.) Arcnet, which runs at 2.5 Mbps allows 2000 ft (610 m) for each coax leg with an active hub. For comparison, twisted pair wiring is limited to 100 m (328 ft).

The standard coax dielectrics are polyethylene and PTFE or FEP, although several other materials are also used. The loss characteristics of the dielectric material contribute to the loss at any given frequency. Low-loss dielectric materials and constructions, such as foam or helical dielectrics are available. However, these low-loss cables are cost-prohibitive for normal LAN cable installations.

Insulation

The insulation in coax cables includes the jacket as well as the insulating dielectric that separates the center conductor and the shield.

The primary types of insulation materials used with coax are polyethylene, polypropylene, various synthetic rubber compounds, polyvinyl chloride (PVC), and fluorocarbon polymers such as polytetrafluoroethylene (PTFE), fluorinated ethylene-propylene (FEP), and ethylene-chlorotrifluoroethylene (ECTFE or HALAR®). Nonplenum coax cables generally use polyethylene as the dielectric that surrounds the center conductor and use noncontaminating synthetic resin for the outer jacket. Plenum-rated coax cables use a PTFE dielectric (such as Teflon®) or FEP and a similar outer jacket.

Color-coding and marking

The most common color for nonplenum coax cable is black. Most of the nonplenum cable that is used for LAN wiring will have a black synthetic rubber or PVC outer jacket with white markings. Plenum cable usually has a white or translucent jacket with black markings. A notable exception to this rule is

TABLE 3.7 Types of Coax Used in LAN Wiring

Coax type	Impedance	LAN use	Special considerations
RG-6/U	75 ohms	Arcnet	Smaller and cheaper alternative to RG-11/U. Lower loss but more rigid than RG-59/U. Must use RG-6/U connectors.
RG-8/U	50 ohms	Ethernet Thicknet	May substitute for 802.3 cable in an emergency or special circumstances.
Rg-11/U	75 ohms	Arcnet	Heavy cable is difficult to work with. See RG-59/U
RG-58A/U	50 ohms	Ethernet	Stranded center conductor required. Original "cheapernet" cable.
RG-59/U	75 ohms	Arcnet	Smaller and cheaper than RG-11/U, this cable is easier to install and readily available, as it is also used for CATV.
RG-62	93 ohms	Arcnet, IBM 3270	The famous IBM Terminal cable can be adapted to Arcnet.
Ethernet 10Base2	50 ohms	Ethernet Thinnet	Cable is marked for 802.3. Double-shielded cable similar to RG-8/U.
Ethernet 10Base5	50 ohms	Ethernet Thicknet	Cable is marked for 802.3. Similar to RG-58A/U with stranded center conductor.

classic "thick" Ethernet (10Base5) cable. Thick Ethernet cable is usually a very bright color such as yellow.

Coax cable should be marked in a similar manner as twisted pair. The cable will generally have a manufacturer's name and part number, the UL/NEC class or use marking, and the appropriate "RG" number. It may also have EIA/TIA markings or IEEE 802.3 markings. Coax use ratings are covered in NEC Articles 800 and 820. If you use RG-58 type cable for thinnet, you should be aware that some of this type of cable does not meet 802.3 specifications and may not be usable at the maximum length for thinnet. You should be sure to specify RG-58A/U that is certified for 802.3 use. EIA/TIA 568 1991 contains specifications for coax, and those markings may be on the cable jacket.

Ethernet trunk cable (thicknet) will have marking bands every 2.5 m to indicate locations for transceiver tap installation.

Shielding

A concentric shield conductor surrounds the center conductor and dielectric insulation in a coax cable. This outer shielding layer causes the electrical and magnetic fields to be contained within the cable and shields the center conductor from fields outside the cable.

Two types of shielding are generally used in coax cables, foil and braid. The foil shield, sometimes called an *overall foil shield*, is a very thin metallic foil that totally surrounds the dielectric. The braid shield consists of a wire mesh of very fine wire conductors. As with the foil shield, the braid surrounds the dielectric and forms a shield. The braid is effective at LAN frequencies even though there are small gaps between the wires. The gaps are not significant compared to the wavelength of the LAN frequencies. Some coax, such as Ethernet trunk cable, uses a double-braid shield to further increase the shielding properties. Cables with braid shields are generally more expensive than cable with foil shields.

Nonpaired Cable

We will cover nonpaired cable in this section to help you identify it. Nonpaired cable is not proper for use as LAN wiring because it lacks the self-shielding properties of twisted pair wire. It both generates and is susceptible to interference, which means that it will probably violate the allowable emission standards of national and international organizations (with government entities such as the U.S. Federal Communications Commission (FCC) this is also a legal violation). Non-paired cable also may not exhibit a consistent characteristic impedance and may create crosstalk between the pairs. This crosstalk happens even at voice frequencies, as you may know if you have two phone lines in your house and have older, nonpaired wiring. At LAN frequencies, the crosstalk between the transmit and receive pairs may mean that you will receive enough of your transmitted signal to cause an error.

You will undoubtedly encounter those that have used short lengths of nonpaired cable (such as the flat telephone cord that is used with RJ-11 connectors) with no apparent problems. Don't believe them. In all likelihood, they are actually experiencing undetected data errors that just slow down their network. In any event, such cable is a problem waiting to happen. What works at 2.5 MHz may be marginal at 10 MHz, impossible at 16 MHz, and a joke at 100 MHz. Don't let nonpaired cable be used anywhere in your network!

General construction

Nonpaired wire may be divided into two general classifications, round cable and flat cable. Both consist of two or more insulated wire conductors and may have a covering jacket over the wires. As with paired cable, nonpaired cable may be shielded with either a foil or a braid conductor.

Round cable, as the name implies, consists of two or more wires that are contained within a round protective jacket. An example of round nonpaired wire is shown in Fig. 3.6. Three or more wires may sometimes be wound in a loose spiral within the jacket, but this is not the paired twisting that provides self-shielding. Round cable may have either stranded or solid conductors. The common variations are the 4-wire solid conductor cable used for inside residential wiring (called "4-wire," "inside wire," "IW," or "JK") and the multiconductor cable used for RS-232 data wiring. The 4-wire cable is famous for causing crosstalk when two phone lines are installed on the same cable. If you are lucky, and your house is newer, you will have telephone grade 4 pair wire. The older RS-232 terminal wiring could use nonpaired wire because the data rate was very low (9600 bps vs. 10,000,000 bps for Ethernet!).

Flat cable is available in two common types: ribbon cable and flat jacketed cable. Ribbon cable is a jacketless design where the insulation of each wire is joined between the individual insulated, stranded conductors. Figure 3.7 shows an example of ribbon cable. The wires are laid side by side, forming a flat cable. The ribbon cable may have an overlaying shield and outer jacket, although this is unusual. Ribbon cables are

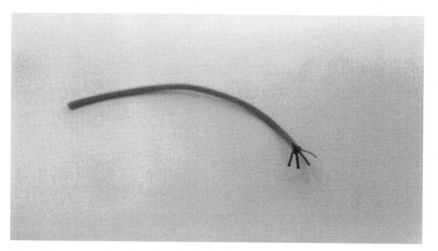

Figure 3.6 An example of round nonpaired wire.

often used for connections to circuit boards within equipment. They generally are not used externally, because the lack of shielding exceeds FCC emission standards.

An example of flat, jacketed cable is the common telephone cord used with modular RJ-11 (style) and RJ-45 (style) connectorized equipment, as shown in Figure 3.8. It is sometimes called "silver satin" or "telephone zip cord," although it is available in many colors and really doesn't zip apart to separate the conductors. Whatever you call it, it consists of two or more insulated, stranded conductors that are positioned side by side and covered with a jacket to form a flat cable. The conductors are run in parallel to each other and are not in any way twisted or paired. Silver satin can also cause crosstalk between pairs, although the placement of the pairs in the cable can actually provide some degree of shielding (for example, pair 1-2 and pair 7-8).

There is also a type of flat cable which consists of paired wires positioned side to side. This cable, which is used for under-carpet applications, has similar characteristics to other paired cables covered earlier in this chapter.

Wire sizes

There are many wire sizes associated with nonpaired wire. Theoretically, any conductor size may be used, but you will probably find that sizes between 18 and 26 gauge (AWG) are the most common for this type of wire. Telephone 4-wire and silver satin are generally 22 and 26 gauge, respectively. Ribbon cable can really be any wire gauge, although the usual range is 26 to 30 gauge.

Electrical characteristics

Nonpaired wire is generally characterized in terms of resistance per linear foot. For example, 24 gauge annealed copper wire has a resistance of 25.7 ohms per

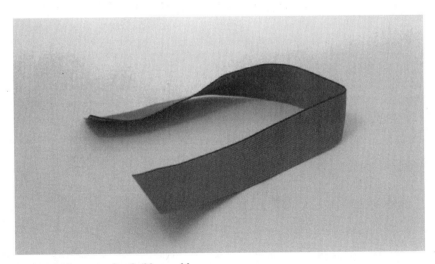

Figure 3.7 An example of ribbon cable.

Figure 3.8 An example of flat-jacketed cable is the common telephone cord that is used with modular RJ-11 (style) and RJ-45 (style) connectorized equipment.

1000 ft. Complete specifications for copper wire are available in several reference handbooks. This is very useful information if you are using the wire to supply power or signaling voltages, such as door bells or fire alarms.

Nonpaired wire is really not intended for transmission at any frequencies above voice, and even there, the transmission characteristics are poor to the extent they are consistent. Of course capacitance may be specified for ribbon cable, since it is used to interconnect electronic circuits.

Insulation

The insulation types for nonpaired wire are identical to those for twisted pair wire. The same insulation types are used and care must be taken in plenum or riser spaces, in accordance with the NEC and local standards.

Color coding and marking

A very wide variety of wire insulation and jacket colors are available for nonpaired wire. However, for our purposes in telecommunications, there are some common color-coding standards that are useful.

The round 4-wire nonpaired cable that is used in some residential wiring used the colors green, red, yellow, and black. This cable was originally designed for single-circuit use—that is, one phone line. The green and red wires supplied the voice signal (tip and ring), and at one time (a long, long time ago), the yellow supplied ringing voltage and the black, ground. The

4-wire cable continued to be used long after the ringing voltage was moved to the tip and ring and the ground was deleted. For a time in the 1960s, the yellow and black were used to supply power from a wall-mounted transformer to the lamp that lit the dial of the famous "Princess" phone. This was later eliminated as the lamp power was also supplied on the tip and ring wires. When a second phone line was to be added, the yellow and black wires were generally available, and that's where it went. Unfortunately, the nonpaired 4-wire cable can generate serious crosstalk between the pairs, and you may faintly hear the other person when you are on a different line. This may be an advantage for parents, but is a problem for LAN circuits and must be avoided.

Shielding

Nonpaired cables may be shielded with foil or braid shields, as with twisted pair cable. A typical use for a shielded, nonpaired cable is for microphone cable. Even flat cable may be shielded. Such cables are useful in meeting electromagnetic interference (EMI) requirements of regulating agencies.

Fiber Optic Cable

Fiber optic cable may also be used in LAN wiring. It is generally specified for backbone wiring, such as vertical wiring between wiring closets and is not yet as common as twisted pair for horizontal wiring. However, fiber optic cabling is one of the acceptable types of cable in TIA 568-A, and it is quite possible to run most LAN topologies entirely on fiber optic cable.

Fiber optic cable offers several unique advantages over conventional metallic wiring. It neither generates nor is susceptible to electrical or magnetic fields. High bandwidth signals can travel very long distances with little attenuation. The nonconductive fiber can eliminate grounding and bonding problems, with the associated electrical safety hazards. Outdoor-rated fiber optic cables often are completely nonmetallic, which makes them resistant to lightening strikes that plague interbuilding metallic cables.

Fiber optic cable still has a few disadvantages, though. For example, when compared to conventional copper twisted pair wiring, fiber optic cable is more expensive, requires more expensive workstation outlet configurations, cannot transmit DC power to remote devices, and might be more difficult and time-consuming to terminate, in addition to requiring special termination equipment. It is also more difficult to test than lower-performance Category 3 copper, where a simple continuity check almost guarantees operation for low-speed LANs. Optical TDR equipment costs, however, are on a par with advanced Category 5 test sets.

The lack of any metallic conductors presents a surprising problem if a defective fiber optic cable must be located. The lack of a conductor means that inductive cable tracers cannot be used to find the suspect cable in a bundle of other fiber cables. In addition, direct, physical length measurements may be needed to pinpoint the location of the break, even though the distance is easily identified on optical test equipment. It may be less expensive to replace an indoor fiber cable rather than attempt to locate a break or other discontinuity. Outside plant cables will need to have faulty sections spliced or replaced, as it would be cost-prohibitive to replace an entire run. However, a longer than necessary section may need to be replaced to allow for measurement uncertainty if the location of the break is not obvious.

Fiber optic cable is much maligned as it is surprisingly fragile. The truth is that the thin glass fiber (about twice the width of a human hair if the cladding is included) is incredibly flexible and resilient given the brittle expectation for something made of silica. Even so, additional care must be taken in handling the fiber, particularly near the point of termination, where the protective jackets and strength members are removed and the thin buffered fiber is exposed.

The use of fiber optic media for LAN and multimedia communication is becoming more popular as time goes on. However, for most installations, fiber optic cable will primarily be used in applications that exploit its advantages over conventional metallic wiring.

The Fiber Debate

The choice between installing copper wiring and fiber optic cable to the workstation is the subject of much debate—an important issue, particularly in new cabling installations where the cable plant is expected to have a useful life of 15 to 20 years. Few would suggest that anything but fiber should be used for campus wiring or for longer high-speed cable runs between wiring closets. The controversy centers on whether fiber should be used from the wiring closet to the workstation—so-called "fiber to the desktop." Current estimates for horizontal fiber range from 5 to 15% of all installations. Copper wiring installations still dwarf fiber installations, yet there is much disagreement on this issue.

What are the two sides to the debate? The proponents of fiber argue that it offers greater signal bandwidth, runs longer distances, offers future growth, costs nearly the same, and is now easy to install and more rugged than generally thought. Proponents of copper over fiber argue that expanded bandwidth is being planned for copper, that the major networking equipment does not support fiber (unless at much greater cost), that the installed cost of fiber is significantly greater than copper, and that fiber is subject to breaks that are expensive to repair. In general, the most avid fiber supporters are either those companies that offer fiber optic cable, connectors, and related hardware, or they are installers and users who have made a significant investment in fiber optic technology and very naturally defend their decision. Please do not construe this statement to mean that either of these two groups is blatantly wrong but simply that they have a certain bias which leads them to the conclusion that fiber is always the best choice. In some circumstances, fiber cabling may clearly be the best, if not the only reasonable choice. But, in many other situations, the issue is less clear. What should be your choice?

The fiber-vs-copper issue can be reduced to three questions:

- Should fiber be used because of its greater bandwidth and for future migration to higher-speed networking?
- What are the cost implications of installing fiber over copper?
- Are there any special applications that favor fiber over copper?

(continued)

The bandwidth question is a moving target. Technology simply will not stay still, so we might as well give up trying to say with certainty what the future will bring. We can say one thing for certain: future networking applications will use much more data bandwidth than in general use today. Copper twisted pair technologies, particularly with the use of enhanced Category 5 products, easily support 100 Mbps now and products are being developed which will take the OC-12 fiber data rate of 622 Mbps over the same copper cable. Of course, fiber can take that signal over longer distances, but 100 meters may be all that is required in horizontal cabling.

The question of relative cost is easier to approach. Fiber boosters have stated that installed fiber costs only about 20% more than Category 5 copper. This is a little hard to prove, at least with the installation companies and manufacturers we have talked to. Fiber optic cable is indeed much less expensive than it used to be. However, cautious users may wish to use protective innerduct or use more expensive strengthened cable to avoid the possibility of fiber breaks that might occur over the installed life of the cable. In addition, no one would argue that the interface adapters for network equipment and hubs cost more for fiber optic capability. Ignoring any of these installation differences, how much do the installation components and cable cost, at the current time?

Let's look at fiber versus copper termination components first. The traditional drawback to fiber installation has been the amount of time it takes to prepare, cleave, epoxy, and polish each fiber connection. New, quick-termination fiber connector assemblies are available that cut termination time to about one minute per connector. However, these connectors may cost (the end user) about $10 each, so you are trading installation labor for component cost. Adding the cost of the fiber optic outlet box with one duplex 568SC adapter (total cost about $35) to the cost of the two quick-termination connectors yields a fiber termination components cost of about $55. In contrast, a good-quality Category 5 modular connector and plate would cost about $15 to the end user. Installation labor at best would be approximately the same, but the fiber termination requires the purchase of an installation kit at a cost of approximately $1000.

Cable prices bounce around a lot because of periodic shortages in plenum-rated materials, but the ratios remain about the same. If the installation calls for plenum-rated cable, then plenum-rated jacketing and inner materials must be used even for fiber cable. The recent plenum price for two-strand plenum 62.5/125 fiber cordage was 70% higher than for the same manufacturer's Category 5 plenum copper cable. (Another way to look at this is that the copper cost was 40% lower than the fiber.) You can draw your own conclusions from these cost comparisons.

The third question asks if your application requires some of the special characteristics that make fiber shine. For example, in an industrial plant with long distances, lots of electro/magnetic field interference, and potential grounding problems, fiber cabling makes lots of sense. Fiber is immune to electrical interference, eliminates ground loops, and can transmit much longer distances than copper. If your network equipment has existing fiber interfaces or if you require very large data bandwidths, fiber is again the obvious choice for your use.

But what about the future? Will technology make copper cabling facilitcies obsolete in a few years? Should you install an all-fiber facility now so that doesn't happen? What about your current equipment that may not have fiber interfaces? Can you afford new network adapters and hubs that support fiber, or can you even upgrade your existing equipment? Wow, what a mess! How can you decide between the two types of cabling?

If relative expense is not an issue, must one really choose fiber OR copper? Why not install both? Install Category 5 copper for today's applications and fiber for the future. In fact, some telecommunications outlet manufacturers have outlet modules that can easily accommodate both a fiber duplex connection and two or more twisted pair cable jacks. It is certainly less expensive to install both technologies in a new facility than it would be to go back in five or ten years and add fiber. The best advice, if the budget allows, is to put in twin fiber and copper facilities.

General construction

Fiber optic cable has a construction surprisingly similar to some multiconductor wire. One or more buffer-coated optical fibers are enclosed in a protective outer jacket with a fibrous strength member in between. This construction is shown in Fig. 3.9. The buffer is rather like the insulation that surrounds a copper wire in a conventional cable. The strength fibers are added to prevent stretching of the cable, which would fracture the optical fiber.

Optical fiber is made of extruded glass (silica) or plastic, specially formulated to pass light of specific wavelengths with very little loss. Most fiber optic LAN wiring is used for backbone wiring and uses glass fiber. Plastic step-index fiber is available from a few sources, but has traditionally suffered from high attenuation and limited bandwidth, factors that are critical to the deployment of fiber technology. Recent developments in graded-index plastic optical fiber (GIPOF) may offer increased bandwidths in the future, perhaps with lower cost and suitable attenuation when compared to glass fiber. Plastic fiber construction is not yet blessed by the standards for LAN wiring, so the rest of this section will concentrate on the characteristics of standard glass fiber.

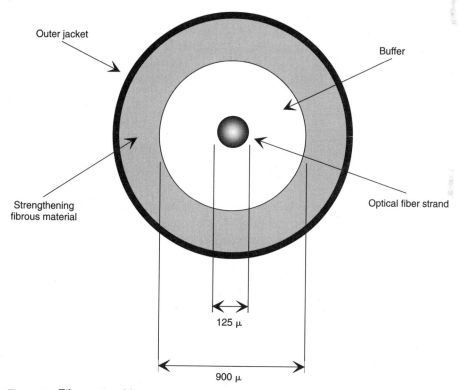

Figure 3.9 Fiber optic cable, crossection.

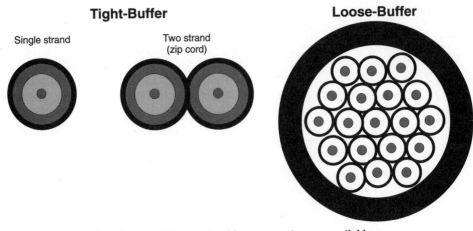

Figure 3.10 Several variations of fiber optic cable construction are available.

Several variations of cable construction are available, as shown in Fig. 3.10. The cable may use either a tight or loose buffer construction. A tight buffer closely surrounds the fiber and adheres to it. The tight buffer must be removed with a tool similar to a wire insulation stripper before the fiber can be terminated in a connector. Loose buffer cables use buffer tubes to surround and protect each fiber, which is loose in the tube. A variation of tight buffer cable is intended for breakout of the fibers without requiring a protective breakout box to safeguard the exposed fibers. This type of cable encloses each fiber in a breakout jacket that contains its own strength fibers. The entire assembly is covered with a protective outer jacket. Filler cords may be used to make the overall cable round. A variation of the breakout cable is the two-fiber duplex, or *zip cord*, fiber cable that is often used for jumper or patch cords.

Fiber optic cables are commonly available with 1 to 36 fibers. At two fibers and above, even numbers of fibers are the most common (e.g., 2, 4, 6, 12) because fibers are often used in pairs for full-duplex circuits. Cables are also available with a combination of optical fibers and metallic pairs. These special-use cables combine the properties of each type of cable, including the advantages and disadvantages.

Fiber sizes

The most common fiber size that is used in LAN wiring is multimode 62.5/125 fiber. This number describes the diameter, in micrometers, of the fiber core and the cladding. The tight buffer surrounds the cladding and brings the overall diameter to about 900 microns. This 62.5/125 is the fiber that is specified in TIA 568-A. It can be used for many applications, including Ethernet 10BaseF, FOIRL, optical Token-Ring repeaters, FDDI, and some implementations of ATM.

Other fiber sizes that are commonly available include 50/125, 85/125, and 100/140 micron fiber. It is important to use the same type and size of fiber in an installation, since transitions between sizes can cause excessive loss. Also, the light-emitting transmitters and receivers are optimized for fiber of a certain size. Using the wrong size fiber will cause some of the light signal to be lost due to poor coupling. This may result in marginal operation or circuit failure.

Optical characteristics

Unlike copper wire, which can carry signals of any frequency from DC to many megahertz, fiber optic cable is designed to carry light in a range of wavelengths. These wavelengths are typically from around 800 to over 1500 nanometers (nm). Operating characteristics of a particular fiber are normally specified at discrete wavelengths which correspond to the output wavelengths of common light-emitting diode (LED) and semiconductor laser emitters/transmitters. Multimode fiber is normally used at 850 nm or at 1300 nm, while single mode fiber is used at 1310 nm or 1550 nm. For any given installation, only one mode and one wavelength are used. The fiber cable should be characterized for operation at that mode and wavelength.

Optical fiber is rather like a waveguide for light. While RF waveguides are hollow with metal sides, optical fiber is solid. During manufacture, the refractive index of the glass is made to vary with the diameter of the fiber core. A difference in refractive index is what makes light actually bend between air and water, for example. The light is sent down the fiber and literally bounces off the "walls" formed by the step in the refractive index. In a single mode fiber, the transmitting emitter is a laser diode, chosen so that the light entering the fiber will be coherent and straight along the axis of the fiber. A multimode fiber, on the other hand, often uses a graded refractive index. Light entering from a light-emitting diode (not a laser) enters at all angles, but is coached down the fiber by the gradient of refraction. Multimode fiber is also available in stepped index, but the step occurs much nearer the outer diameter than with single mode fiber. Single and multimode operations are shown in Fig. 3.11.

Optical fiber can carry very large bandwidths of information at low attenuation. Multimode fiber can be used at distances exceeding 3000 m, although other constraints may limit operation to less distance. Single mode fiber can be used at many times that distance and is frequently used in long distance distribution of T-carrier and Sonet signals. Typical attenuation for 62.5/125 multimode fiber is 3.5 dB/km at 850 nm and 1.0 dB/km at 1300 nm. Typical numbers for single mode 125 micron fiber are 0.5 dB/km at 1310 nm and 0.4 dB/km at 1550 nm.

Transmission bandwidth of fiber optic links are primarily limited by the bandwidths of the transmit and receiving devices, rather than the fiber core's transmissivity. However, as you can see from the attenuation characteristics above, the attenuation does vary somewhat with wavelength of the transmitted light. Bandwidth is also limited by chromatic dispersion of the transmitted signal, which increases linearly with distance along the fiber. Short distance transmission, such as in the horizontal link, greatly minimizes this effect. When used for low bandwidth LANs, such as Ethernet 10BaseF, the bandwidth of the link is

62.5/125 μ Multi mode fiber-graded index

62.5/125 μ Multi mode fiber-step index

125 μ Single mode fiber

Figure 3.11 Single and multimode fiber optics operation.

rarely a problem. At 100 to 150 Mbps, though, bandwidth of the link must be maxmized to several times the signaling rate in order to maintain acceptable received signal fidelity. For example, FDDI bandwidth should be 500 MHz.

Insulation/Jacketing

Insulation/jacketing material types are roughly the same for fiber optic cable as for metallic cable, but the rating codes are different. The ratings include plenum use, general purpose, and conductive cable types. Conductive fiber optic cables contain non-current carrying metallic strength or vapor barrier components. A listing of the fiber optic cable types is shown in Table 3.8. For more information, see Section 770 of the NEC.

From a flame-resistance standpoint, the requirements for the use of fiber optic cable are identical to conventional cable. Plenum-rated fiber cable must be used in air plenums and riser-rated cable must be used in vertical riser shafts. In general, a plenum-rated cable may be substituted for riser use or general use. Likewise, a riser-rated cable may be used for general use. Ports and pathways for fiber optic cables should meet the appropriate flammability and fire-stop requirements, as with metallic cable.

Fiber optic cables that are intended for outside use or for direct burial should be appropriately rated by the manufacturer. Because fiber optic cables do not carry voltages, the cables technically do not need to meet the NEC requirements for direct burial; however, the resistance to moisture, chemicals, and abrasion should be the same. Remember that a cable that runs any length inside a plenum must be appropriately rated, even if that cable originates outside the plenum space. An exception to this rule is an outside cable at an entrance facility. See the NEC for details. Fiber optic cable with UV resistant jackets and armoring is also available.

Color coding and marking

Fiber optic cables are generally colored black for outdoor use. For indoor use, a variety of colors are available. You may wish to choose a bright color to distinguish the fiber optic cable from other cable. Fiber optic jumpers are often bright orange or yellow. The glass fiber may be damaged by tight bending or sudden impacts and is easier to avoid when it is brightly colored.

Marking is the same for fiber optic cable as for conventional metallic cable. The marking will generally include the name of the manufacturer, the manufacturer's

TABLE 3.8 National Electrical Code (NEC) Cable-use Codes for Fiber Optic Cable

NEC article	Code	Meaning	Allowable Substitutions*
770	OFNP	Optical Fiber Plenum Nonconductive	none
	OFNR	Optical Fiber Riser Nonconductive	OFNP
	OFNG OFN	Optical Fiber Nonconductive General purpose	OFNR
	OFCP	Optical Fiber Plenum Conductive	OFNP
	OFCR	Optical Fiber Riser Conductive	OFCP OFNR
	OFCG OFC	Optical Fiber Conductive General purpose	OFNR OFCR OFN

*NOTE: In general, a cable with a more strict usage code may be substituted in an application that allows a less strict usage code. For example, a cable rated to the stricter Plenum code may be substituted in a Riser application. The chart does not list all possible substitutions, only those to the next level. A nonconductive cable is one that contains no metallic elements, including conductors, metallic sheaths, or strength members.

SOURCE: *1996 NEC*

part number, the NEC or UL rating (or similar rating for the country of use), and perhaps the fiber size (such as 62.5/125) and mode.

Shielding

Fiber optic cables transmit light and do not need to be shielded from electromagnetic fields. Some cables are armored or include a wire strength member for aerial installation. You should be aware that the metallic armoring or wire reduces the cable's relative immunity to lightning. In addition, under some circumstances, high levels of static electrical charge can build up on outdoor fiber optic cables, and workers may need to use special precautions including grounding of the cable to discharge built-up voltages.

Combination cables are available that contain both optical fibers and metallic conductors. These cables may be offered with shielding for the wires. See the section on UTP and STP cable for a complete discussion of shielding types for these conductors.

Special-Purpose Wire

If you intend to place LAN cable outside of ordinary office locations, you may need to specify special purpose cable types. Most normal locations for LAN cable are benign environments that maintain the cable at mild temperatures, away from moisture, contaminants, harsh or dangerous chemicals, and the effects of the sun. However, many locations exist that require special wire insulation, jacketing, armoring, or ratings. Most wire types are available in special versions to accommodate all of these needs. Several of these special-use ratings for cables are described here.

Plenum and riser rating

Cables that are placed in plenums must be plenum rated. A plenum is a space used to move air to workspaces for the purpose of ventilation, heating, or cooling (HVAC systems). The informal words for plenums are "air duct" and "air return."

In small buildings, plenums can be individually ducted for forced air and return air. Such plenums rarely contain cabling, and you may be able to use cheaper non-plenum rated cable (if you stay out of the plenums). Large buildings often have individual forced-air ducts to deliver conditioned air to the workspace, but use vents into the over-ceiling area to return "used" air to the air handler. In such cases, the entire area above the ceiling grid is classified as plenum space.

The whole purpose of plenum rating cable is to reduce or eliminate the transport of hazardous gasses to uninvolved workspaces, in the event of a fire. Cable placed in the plenum spaces must be tested for flammability and smoke generation to minimize this hazard.

Riser cable is intended for use in vertical shafts that run between floors. Many buildings have a series of equipment rooms that are placed vertically in a reinforced shaft for the purpose of enclosing power distribution equipment,

HVAC units, telephone distribution, and other utility services throughout the building. Large buildings may even have multiple utility shafts. Cable placed in these shafts must not contribute to the spreading of fire from floor to floor. However, the smoke generation requirements are not as important, because the shafts are not plenums. For that reason, the riser rating is less stringent than plenum. Plenum rated cable may generally be substituted for uses requiring riser rating.

These special plenum and riser cable requirements are covered in detail in the NFPA's National Electric Code. NEC Sections 770 and 800 explain these ratings for fiber optic, LAN, and communications cabling.

UV light rating

The ultraviolet light from the sun can badly damage cable jackets that are not specially formulated. This damage will not be immediate but will take place over a period of years. The cable jacket will become discolored, then cracked and brittle. It will fail to keep out moisture and contaminants. Moisture penetration may cause the LAN connection to fail.

UV-rated cable may be obtained from most cable manufacturers. In fact, many cables already include sunlight resistance as a standard feature. You should check with your wire manufacturer if you are not certain.

Outside locations

Cables must be specially designed for use outside a building. The harsh outside environment of temperature, moisture, light, and stress place unusually stringent requirements on cable of any type. Outside cable is suspended long aerial distances, run in underground cable ducts, and directly buried. Cable may need to be gel-filled or pressurized to exclude moisture. It may need special strength members or an integral messenger cable to span aerial distances. Metallic armoring may be needed to resist damage from animals.

If you must run outside cables into buildings, you may want to consider a cable that is also rated for indoor use. Such cable meets the NEC code requirements for indoor use as well as being designed for the harsher outside environment. The NEC specifies that typical outdoor cable may only be extended 50 feet inside a building, unless that cable is rated for the proposed indoor use. The alternative is to place a junction or splice point inside the building, near the point of cable entry, to convert to inside-rated cable.

Hazardous locations

Special cable, conduit, and junction boxes may be required for cable that is run in so-called hazardous locations, including places where flammable or explosive gasses or liquids might be present. Surprisingly, hydrogen and oxygen gas fit into this description, as do volatile liquids such as gasoline and solvents. The NEC addresses the installation of cabling in hazardous areas.

4

LAN Wiring
Topologies

An understanding of LAN wiring topology—how the wires are hooked up—is useful in any discussion of LAN wiring. In this book, we cover wiring suitable for many of the popular types of LAN topologies used today. Most of the other chapters are concerned with wiring devices and techniques that are not specific to any particular LAN type. This chapter will detail several LAN types and their variations, with reference to specific wiring techniques, where appropriate.

We will explain the wiring patterns and connections that are needed to support each type of network, which is called its *topology*. We will not be covering the intimate details of a LAN's protocols, as this is not our topic.

If you are familiar with LAN protocols or the OSI (Open Systems Interface) model, you will recognize the LAN topology as Layer 1, the physical layer. Layer 1 is subdivided into additional layers. Part of Layer 1 comprises the actual wire, connectors, and hubs that must be connected properly for the rest of the protocol model to function. This wiring layer is critical to the functioning of the network. If the wiring fails, the network fails. Nothing else in the protocol stack matters. This may be held over the head of the nearest cable installer (perhaps your own head), if it gives you any comfort.

Many types of LAN topologies may be supported on the twisted pair wiring described elsewhere in this book. However, many of these LAN topologies function on other types of wiring. For completeness, we will include LAN wiring methods that use coax and optical fiber, as well as shielded and unshielded twisted pair.

The three basic types of network topologies are bus, ring, and star. Some LANs, such as Ethernet, use more than one basic type. The three topologies are illustrated in Fig. 4.1. Each topology has its advantages and disadvantages. In some cases, such as Token-Ring, one topology can be wired very much like another. (The "ring" is normally cabled like a "star," but the legs are interconnected at the hub in the wiring closet so as to form the ring.) As you can

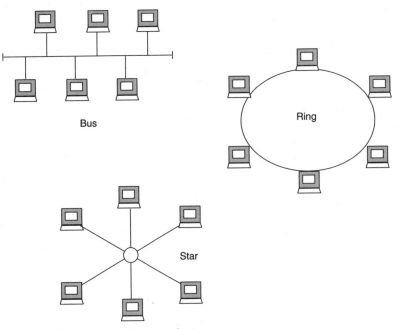

Figure 4.1 The three basic LAN topologies.

imagine, each topology and LAN standard has its fans, as well as its detractors. We will stay away from the contest, except to point out when a wiring method can be used for several topologies. We would assume this to be an advantage for the long term.

Ethernet Coax

Ethernet topology is the granddaddy of LAN systems. It was the first widely accepted, nonproprietary, standardized multiple-access network. Intended for the routine connection of computers and workstations, it allowed many devices to share a high-speed communications line without regard to the customary hierarchy of permanent computer-to-terminal connections.

Ethernet was originally developed by the Xerox Corporation at its Palo Alto Research Center. In the late 1970s, Xerox Corporation, Digital Equipment Corporation (DEC), and Intel Corporation agreed to jointly promote Ethernet as an open standard for computing. This standard eventually became IEEE 802.3, administered by the Institute of Electrical and Electronic Engineers. IEEE 802.3 has been revised and refined several times and now includes coax, twisted pair, fiber optics, and even wireless connection methods. Modern Ethernet, as embodied in IEEE 802.3, actually differs in some ways from the original Xerox/DEC/Intel standard (sometimes called DIX). This older standard has been revised and lives on as Ethernet Version 2. However, most installations have shifted to the IEEE standard and we will refer to that simply as "Ethernet" in accordance with common usage, unless there is a significant difference.

Ethernet uses a method of signaling called Carrier Sense Multiple Access with Collision-Detect, (CSMA/CD). The signaling rate is 10 Mbps and uses a special method, called *Manchester coding*, to preserve a null DC level on the cable; this allows collisions to be detected when they occur. Each device listens to the network for transmissions from other devices. If a transmission is received that matches the device's address, it is processed in accordance with higher-level protocols. If not, the transmission is simply ignored. When a device wishes to transmit, it first "listens" for another existing transmission. If none is present, it transmits its message, and then notes the voltage level on the line to see if another device happened to transmit at the same time. This would be called a *collision*. If a collision occurs, each device waits a short, random time period and retransmits, after checking for an existing transmission. Thus, many devices have access to the network (multiple access), they each listen before transmitting (carrier sense), and they retransmit if a collision occurs (collision detect).

Thicknet (10Base5) cabling

The original implementation of Ethernet used a large, 50 ohm coax trunk cable that is now referred to as thicknet. This cable is specified in the 10Base5 standard of IEEE 802.3.

A 10Base5 network consists of a thicknet backbone cable that is tapped with a series of "transceivers" or "media attachment units" (MAUs). Each workstation or server is connected to a single transceiver with a transceiver cable. Thicknet is a Tapped Bus topology. A typical thicknet installation is shown in Fig. 4.2. The transceiver literally taps directly into the Ethernet cable by

Figure 4.2 A typical thicknet 10Base5 installation showing distance limitations: 500 m segment, 2.5 m tap-to-tap, 50m AUI.

means of a drilled hole and a probe contact. The transceiver has a 15-pin connector, which carries the Ethernet 10 Mbps LAN signal, error and status signals, and power. This interface is called the Attachment Unit Interface (AUI). The transceiver is connected by a transceiver cable, often called an AUI cable, to the workstation or server. This AUI cable is a stranded, 20 gauge, 4 twisted-pair, shielded cable terminated at each end by a 15 pin d-shell connector with a special locking mount. The AUI connections are shown in Fig. 4.3.

The thicknet cable is run, in a continuous fashion, from a starting point to an area near each workstation or server that is to be connected. If a cable is cut, or if another length of cable is to be attached, both cable ends are connectorized and a barrel connector is used to join the two sections. Type-N connectors are used with thicknet. The backbone cable must be terminated at each end by a 50 ohm terminating resistor, which is usually incorporated into a coax connector. The terminating resistors minimize reflections of the 10 Mbps signal that would otherwise occur. Removing a terminator or cutting a backbone cable will cause the network to fail, as the reflections are inverted and cause a failure in the carrier sense and collision detection mechanism. Transceivers may generally be connected to a live network cable, although there is always a possibility that the cable shield and center conductor may short briefly during the process. Such a short will temporarily disable the network, but should be cleared quickly when the tap is installed. Some network managers schedule transceiver installations only during scheduled downtime.

Transceivers are mounted directly to the thick cable and secured in place. A stepped hole is drilled into the cable with a special drill and guide. When the transceiver mounts to the cable, a probe makes pressure contact with the cable's center conductor and the ground of the transceiver is placed in contact with the shield of the cable. (As mentioned, this connection method is sometimes called a "vampire tap.")

Thicknet is most often placed into ceilings or walls near the workstations. Prewire methods may place the cable into wall outlets with an integral

1

Transceiver connector
male DB-15 style

15

Pin	Function-IEEE 802.3	Ethernet V2.0
1	Control-in shield	Shield
2	Control-in A	Collision pres.+
3	Data-out A	Transmit +
4	Data-in shield	−
5	Data-in A	Receive +
6	Voltage common	Power retn.
7	Control-out A	−
8	Control-out shield	−
9	Control-in B	Collision pres.−
10	Data-out B	Transmit −
11	Data-out shield	−
12	Data-in B	Receive −
13	Voltage	Power
14	Voltage shield	−
15	Control-out B	−
Shell	Protective Ground	

Figure 4.3 The AUI connections.

transceiver tap. Power for the transceiver is provided from the workstation when the transceiver cable is connected to the wall outlet. The cable may also be placed above the ceiling, suspended from structural members or lying directly on the ceiling grid. The transceivers are tapped into the cable as needed and the transceiver cables are run to the workstations by dropping directly or running through a wall opening.

The backbone trunk cable may be run a maximum of 500 meters (about 1640 feet) without using repeaters or bridges, with each of these runs being called a *segment*. Segments may be joined with a repeater or a bridge. A repeater simply receives, reconstitutes, and retransmits the Ethernet signal (packet) to the next segment. A bridge or router, on the other hand, reads the packet and determines if the packet's destination is in another segment. If so, the bridge or router recreates the packet and retransmits it on the next segment. Ethernet 802.3 has a maximum limit of two repeaters between segments (four repeaters, if two of the segments are repeater-to-repeater links). Ethernet version 2 defines things a little differently to determine the repeater limit. Bridges and routers may be used to extend the network beyond this repeater limit or to off-premise locations via wide-area networking (WAN).

The transceiver cable is limited to a length of 50 meters (164 feet). While this distance might mean that it would be possible to run all the way from a wiring closet to a workstation, the usual procedure is to run the cable to a nearby section of the trunk cable and install a transceiver. Each transceiver must be separated from others by a minimum of 2.5 m (about 8 feet, 2.5 inches). This minimum distance makes it impractical to place many transceiver taps in a single location. In instances where several AUI cables must be run, a multiport transceiver with several AUI interfaces may be used. The 10Base5 standard allows a maximum of 100 transceiver nodes per segment, including repeaters.

Thicknet installations are most often found in large buildings and are frequently older network installations. The problems that may occur are often due to corrosion of the connectors, barrels, or even transceiver probes. Additional problems may occur due to improper grounding. More information on grounding is given at the end of this section.

Thinnet (10Base2) cabling

A newer implementation of Ethernet is the 10Base2 thinnet standard. This cabling uses a less expensive cable type that is basically RG-58A/U 50ohm coax cable. Thinnet was developed during the early 1980s as a cheaper, easier-to-install version of traditional Ethernet. Sometimes the names thin Ethernet and cheapernet are used to describe 10Base2. Thinnet is also a Tapped Bus topology. A typical thinnet installation is shown in Fig. 4.4. This cable is specified in the 10Base2 standard of IEEE 802.3.

The main topology difference between thick and thin Ethernet is the disappearance of the discrete transceiver and its associated AUI cable. In thinnet, the transceiver is an integral part of the network adapter of each workstation

Figure 4.4 A typical thinnet 10Base2 installation showing distance limitations.

or server. The thinnet coax is simply routed from one workstation or server to another in a daisy-chain fashion. At each workstation's network adapter, a T connector is used to effectively "tap" the coax. In general, a coax cable runs to each workstation where it is connected to another coax cable that runs on to the next workstation, in turn, until all have been connected. At the two ends of the run, a 50 ohm terminator is placed to minimize reflections of the LAN signal. As with thicknet, removing a terminator or disconnecting a cable will cause the network to fail.

The connectors that are used with thinnet are called *BNC connectors* (see Fig. 4.5). The cable-end BNC connectors are a male, bayonet-style (push on and twist to lock) connector, originally used for RF connections. The BNC-T has two female BNC connectors at the top of the T, which are connected to each incoming cable or a terminator, and one male BNC at the bottom of the T, for connection to the workstation adapter.

Thinnet is often placed into the walls or simply run along the floor and behind furniture to connect to workstations. If a wall plate is used, two BNC bulkhead connectors must be provided, since the cable must be extended all the way to the workstation and then back to the wall to complete the network path. For grounding reasons, isolated-bulkhead connectors should be used (see the section on grounding, below). If no workstation is present, a short length of coax, with connectors at both ends, can be used to jumper the two bulkhead connectors and maintain the network path. Sometimes, rather than placing an outlet at the wall, the two lengths of coax cable are run unbroken through an opening in the wall, or simply dropped down from the ceiling. In this case, the two cables are simply joined with a BNC-T, whether or not a workstation is present.

An entire thinnet segment is limited to 185 m (607 feet). This includes every leg as well as the small contribution of the connectors and T's. Although each consecutive length of cable to each workstation is added to the total, thinnet still makes conservative use of cable. Adjacent offices can be wired together with only a small run of cable. Workstations in the same room can be served with very short runs of cable (0.45 m or about 1.5 feet minimum). Thinnet may have as many as 30 nodes on a segment. As with thicknet, only two repeaters may be used between any two workstations in a network, without using bridges or routers (four repeaters, if two of the segments are repeater-to-repeater links). This means that no more than three 185 m workstation segments can be linked with repeaters. Bridges and routers may be used to extend the network beyond this limit or to off-premise locations via wide-area networking.

Problems to watch for in thinnet installations are intermittent connectors and T's, connectors which have pulled loose from the cable, and easily disconnected BNC-T's. The T may be disconnected from the workstation without disrupting the network, as long as both cables to the next adjacent workstations remain connected to the T. In large installations, ground loop currents between areas of separate electrical service may cause network errors and even safety problems.

Grounding

Proper grounding of coaxial LAN cables is very important, both for safety and operational reasons. Both types of Ethernet coax use unbalanced "ground-referenced" signaling. That is, the LAN signal is applied to the center conductor and is referenced to a signal ground, connected to the shield. Most people assume that a ground is a ground. However, it is intended that this signal ground be independent of the "earth" ground (sometimes called *chassis* or *safety ground*). Unfortunately, the independent signal ground of the network and workstations can sometimes get interconnected with safety grounds. If the signal ground on a cable and the chassis ground are not at the same electrical

Figure 4.5 The connectors used with thinnet are called BNC connectors—BNC male cable connectors and BNC-T's.

potential, significant signal distortion and even potentially lethal voltage differences can result. Grounding is illustrated in Fig. 4.6.

The Ethernet specification calls for the trunk cable to be grounded at each wiring closet or thinnet cable to be connected to earth ground at one end (not at both ends). In addition, you should insure that the building electrical system is properly bonded and grounded together, preferably at the one point where the power entry to the building is located. All electrical distribution rooms in the building should be grounded back to the same entry point, as should all power transformers. This is clearly an area for an experienced electrician, but as the network manager or contractor, you should know how to assess a proper building ground system. EIA/TIA 607 gives extensive information on providing proper building grounding and bonding.

Ethernet Twisted Pair

The introduction of twisted pair wiring into standard Ethernet networking ushered in a new age of network connectivity. For some time, frustrated users

Figure 4.6 Typical grounding recommendations for coaxial Ethernet cabling.

had been experimenting with thinnet baluns in order to use existing telephone wire for network links. Several manufacturers even introduced proprietary network adapters and hubs for twisted pair Ethernet. Eventually a standard was fashioned, under the umbrella of the IEEE 802.3, to deploy a new twisted pair Ethernet topology called 10BaseT.

This new standard has been well accepted and robust, indeed. It has been so successful that it has virtually eliminated the installation of new 10Base5 and 10Base2 networks. The need to specify a system of universal telecommunications cabling that would allow the proper operation of 10BaseT gave rise to a series of new cable, wiring, and component standards that eventually resulted in the TIA 568-A standard.

Now that 10 Mbps Ethernet is beginning to seem a little slow, Ethernet networking is taking a new jump to 100 Mbps over twisted pair cable. Fortunately, the standards activities have made much progress and have released 100 Mbps standards for network adapters, hubs, cable, and wiring components.

10BaseT networking

The 10BaseT Ethernet topology is an Active Star, rather than the Tapped Bus topology of 10Base5 and 10Base2. This star topology is quite compatible with the standard home-run method of commercial telephone-style wiring. Although the use of existing telephone wiring was originally the goal of a twisted-pair Ethernet, the standards now recommend that LAN twisted pair wire be used exclusively for the LAN network connection. Telephone wiring should be done using a separate cable to be totally in step with the standard.

A typical 10BaseT installation is shown in Fig. 4.7. The center of the star topology is a 10BaseT hub. Each workstation (or server) has a 10BaseT network adapter port that is connected to the hub over a twisted pair cable. The standard specifies a maximum 90-meter distance for each cable leg, plus a total of 10 meters for interconnection, both at the workstation and at the hub. Modular cords, jacks, punchdowns, cross-connects, and patches are all allowed. While these devices resemble their telephone counterparts, there are important differences that make them "data grade." Some of these differences are evident by the performance certifications needed to meet the standards for LAN wiring.

A significant difference between 10BaseT and other coax Ethernet topologies is the addition of an active hub device. 10BaseT still uses the CSMA/CD signaling method. With the coax topologies, any transmission by a station is passively distributed by the coax cable to all the other connected stations. In 10BaseT, however, a transmission by a workstation first goes to the hub, which then repeats (retransmits) the signal to all of the other connected stations (see Fig. 4.8).

A 10BaseT hub typically has 8, 12, 16, or even more 10BaseT ports. It may be a standalone hub, or part of a chassis with plug-in hub cards. Standalone

Figure 4.7 A typical 10BaseT installation.

hubs usually have one port each for thicknet and thinnet connections, in addition to the 10BaseT ports. The thicknet port is an AUI interface and the thinnet port is a BNC connector. Hubs may be used in concert to build bigger networks by interconnection through the thicknet port, the thinnet port, or one of the 10BaseT ports. The 10BaseT interfaces can be converted to AUI or thinnet interfaces or even fiber optic links with appropriate transceivers or repeaters.

The hub gives 10BaseT some unique advantages over the coax topologies. Because each hub port is repeated to the rest of the network, each 10BaseT port is independent of the others. This means that the cable length of a given port is not affected by the cable length of any other port. It is much simpler to plan and test a 10BaseT network, because the cable length for a port must merely be less than the allowable maximum of 100 m (90 m for the horizontal cable, and 10 m for the patch and user cords at either end). As an added benefit, the hub can automatically isolate any one port that misbehaves.

Workstations may be connected to or disconnected from 10BaseT hubs without interrupting other stations on the network. Hubs usually have status lights that indicate proper connection on the 10BaseT ports and collisions or other error conditions on the thin- and thicknet ports. Standard 10BaseT hubs will automatically isolate (or partition) a port for a wiring reversal, short, or open connection (including no device attached). More sophisticated hubs may also detect more subtle errors and allow for automatic or manual port isolation.

Hubs may also be connected to a variety of management tools, from Simple Network Management Protocol (SNMP) to sophisticated network hardware managers, such as Hewlett-Packard Openview and Sun NetManager.

The basic interface for 10BaseT is the *medium dependent interface* (MDI), a special wiring configuration of the common 8-pin modular jack that is used by some telephone equipment. Two pairs of wires are used, one for the transmit data, and one for the receive data. The connections are polarity sensitive. That means that the connection will not operate if the two wires of either pair are reversed. The wiring of the 10BaseT interface is shown in Fig. 4.9.

Although 10BaseT only uses two wire pairs, it is customary to use 4 pair cable to make the station drops. Several of the 100 Mbps network schemes require all four pairs. In addition, a jack that is wired for all four pairs can support many other types of data and voice connections, which may use some other combination of the eight connector pins. Some installations use the other two pairs for a telephone connection or for another 10BaseT connection. Be wary of telephone installers "stealing" your network wiring pairs for their use. Telephone wiring is often at cross purposes from network wiring. The standards are different, as are the installation practices.

The TIA 568-A wiring standard supports 10BaseT wiring. The wiring standards of Category 3 are normally adequate, although you may wish to use Category 5 in a new installation to allow for future enhancements.Cabling for 10BaseT is simply a wiring star of hub and nodes. Generally, a hub is located centrally, perhaps in a wiring closet, and station cables are run to

Figure 4.8 In 10BaseT, a transmission by a station first goes to the hub, which then repeats (retransmits) the signal to all of the other connected stations.

each workstation location. The cables may be terminated in a patch panel or on a punchdown block adjacent to the hub. A patch cord or octopus (fan-out) cable connects each station cable to a hub port. These terminations are covered in detail in Chapter 8.

At the workstation location, station cables are normally terminated in a wall jack and a user cord connected to the workstation. The wiring is the standard 10BaseT wiring shown in Fig. 4.9; however, the more robust wiring pattern of TIA 568-A offers effectively the same connections. Connector jack internal wiring may present some confusing variations in wire colors and numbering; some of these variations are shown in Fig. 4.10. Note that the wiring colors may be completely different than those of the station cable and the plates sometimes have numeric markings that do not correspond to the pin numbering of the jack itself. Fortunately, 10BaseT has the ability to sense misconnections, and the hub status light for the port is a good diagnostic tool. Inexpensive modular test indicators are also available to help you make this connection.

All wiring in 10BaseT is straight through (pin 1 to pin 1, pin 2 to pin 2, and so on). Be careful to use only adapters, patch cords, adapter cables, and accessories that are wired this way. Many similar cables and adapters that are intended for telephone use are wired "cross-over", which means pin 1 to pin 8, etc. In addition, all 10BaseT wiring is twisted pair. Avoid using the flat telephone wire, such as silver satin, that is sometimes available with 8-pin modular connectors.

Each leg of a 10BaseT network is limited to 100 meters (328 feet). Of this 100 m, 90 m typically may be station wire and an additional 10 m may be the total for the patch cord, cross-connect (if any), and workstation cable (user cord). Cable, component, and installation standards should be in accordance with TIA 568-A, Category 3 or higher. Note that either T568A or T568B modular jack

Figure 4.9 The wiring of the 10BaseT interface.

Full 4 pair jack wiring
(For one 10BaseT connection)

Station cable Wiring	T568A:	W/G	G/W	W/O	B/W	W/B	O/W	W/Br	Br/W
	T568B:	W/O	O/W	W/G	B/W	W/B	G/W	W/Br	Br/W
Internal jack wiring		B	O	Bk	R	G	Y	Br	W
Jack pins		1	2	3	4	5	6	7	8

"Duplex" wiring
(4 Pair cable split to two 10BaseT jacks)

W/O	O/W	W/B	-	-	B/W	-	-		T568A	W/Br	Br/W	W/G	-	-	G/W	-	-
W/B	B/W	W/O	-	-	O/W	-	-		T568B	W/G	G/W	W/Br	-	-	Br/W	-	-
B	O	Bk	R	G	Y	Br	W			B	O	Bk	R	G	Y	Br	W
1	2	3	4	5	6	7	8			1	2	3	4	5	6	7	8

Jack 1   Jack 2

Legend

W	White	Br	Brown
B	Blue	R	Red
O	Orange	Bk	Black
G	Green	Y	Yellow

Figure 4.10 Typical 10BaseT jack wiring. Some connector jack internal wiring may present some confusing variations in wire colors and numbering. Note that the wiring colors may be completely different than those of the station cable and the plates sometimes have numeric markings that do not correspond to the pin numbering of the jack itself.

wiring patterns will support 10BaseT. Just be sure that you use the same pattern at each end of a cable, whether in the wall or as a user/patch cord. Termination of the wires is covered in detail in Part 2.

Problems in 10BaseT usually relate to individual workstations. If you are having a problem, the hub status lights can diagnose total failures of the cable to a workstation or server. Common cable failures include damaged wires, connectors pulled loose or not properly seated, and poor or improper connections at the modular plug. Use only the expensive, tool and die type, crimp tools for 8-pin plugs. The cheaper tools will not properly seat the center contacts on the plug and the cable may fail. In addition, you should be very cautious if you use solid wire with modular plugs (see Chapter 10 for more information). If the cable tests good, suspect the workstation's network adapter, if the problem is with a single workstation. If all workstations are affected, suspect the hub or any thinnet or thicknet connections between hubs or to other devices.

100BaseT cabling

Standards for 100 Mbps networks have been in a state of rapid development. Many proprietary methods and interim standards have been put forward over the time since 10BaseT was first introduced. The IEEE 802 standards committee has resolved the competing technologies with the introduction of a supplement that details the implementation of 100BaseT. There are currently two copper wire 100BaseT variations under what has become known as the IEEE 802.3u Supplement standard. They are 100BaseTX and 100BaseT4. A companion 100BaseFx (covered later in this chapter) rounds out the 100 Mbps CSMA/CD standard offering. The non-CSMA/CD 100VG-AnyLAN (which we cover in the next section) was relegated to a new committee, IEEE 802.12, after much debate.

In some ways, the new standards encompass several formerly proprietary technologies under the same standards umbrella. The variations are not cross-compatible, although some vendors may offer adapters or hubs that can meet more than one standard. You should still be cautious in what you get, as the product nomenclature can be quite ambiguous.

These 100 Mbps Ethernet topologies are sometimes referred to as Fast Ethernet. The implementations of the two main standards, TX and T4, differ in the minimum link performance category and the number of pairs required. The important point is that either of the copper Fast Ethernet standards can be supported on an unmodified TIA 568-A cabling system, although the required category of wire is different. And implementations of both fortunately offer backward compatibility with 10BaseT signals and networking equipment. Because of the differences, though, let's cover them one at a time.

100BaseTX

The simplest of the Fast Ethernet standards is 100BaseTX. This standard is quite similar to 10BaseT, except that it runs at 100 Mbps, 10 times faster. Also, it requires only two pairs, one for transmit and one for receive. The pairs are wired exactly the same as the 10 Mbps version. The main difference from the slower standard is that the signal itself is at a full 100 Mbps data rate. A Category 5 link is required. The signal is simply too fast for Category 3 or 4 cable. This may be the primary justification for placing

Category 5 standards into any new cable installation. Remember that all components (not just the wire) must meet Category 5 and proper installation techniques must be used.

The wiring pattern for 100BaseTX is shown in Fig. 4.11. Although only two pairs are required for this topology, it is still recommended that all four pairs be connected in the T568A pattern given by TIA 568-A.

An additional feature may allow negotiation between NIC and hub to enable a fall-back to 10BaseT if both devices are not 100BaseT compatible. This is a real advantage if you are gradually upgrading to 100 Mbps. You can install dual 10/100 Mbps NIC cards in all your new workstations at relatively little extra cost. When you change your hubs to 100BaseTX, the dual speed cards will automatically make the 10 times speed shift.

A 100BaseTX network has some rather severe distance restrictions. Because of the timing constraints of CSMA/CD, the total "radius" of the network is a tenth the size of a 10 Mbps Ethernet network. Hubs may be linked, but only on a very short backbone of 5 meters. The hubs (which are repeaters) can be linked to stations by 100 m of cable, which fits into the TIA 568-A limit of 90 m + 10 m nicely. This computes to a maximum station-to-station distance of 205 m (100 + 100 + 5). A fiber link from a hub to a bridge, router, or switch can be up to 225 m, and a non-hub fiber link can be 450 m with CSMA/CD enabled and 2 km disabled. These limits may encourage the use of large, multiport hubs and Fast Ethernet switches.

An interesting problem exists if the environment has both 10 and 100 Mbps stations. Any simple hub must be only one speed, either 10 or 100, because a 100 Mbps station obviously could not transmit to a 10 Mbps station. A speed-translating bridge or router would have to be used. You could perhaps do this with a "switch," but then the switch terminology probably obscures the real function.

8-pin Modular jack
(Front view)

Figure 4.11 The wiring pattern for 100BaseTX.

100BaseT4

For networks built to accommodate 10BaseT wire speeds, the move to 100 Mbps is a tough one. Many twisted pair networks have been installed to Category 3 wiring standards that will not accommodate the 100BaseTX line speed. For those installations that have installed 4 pair Category 3 cable, the new 100BaseT4 standard allows the operation of a 100 Mbps CSMA/CD connection over Category 3 cable.

The implications of this are enormous, if you have a large installed base of 10 Mbps (or 16 Mbps) users. You can selectively increase your LAN connections to 100 Mbps without pulling new Category 5 cable. This can save a lot of money if you have a large number of terminals. Better yet, many of the suppliers of 100BaseT4 workstation adapters offer dual speed operation at 10 and 100 Mbps. The adapter automatically detects which speed is in operation and makes its connection at that speed.

However, if you are installing a new cabling system, you should not use the existence of this type of high speed connection as an excuse to put in Category 3 links. Full Category 5 installations are not that much more expensive than new Category 3 ones, and you will be ready for future technologies that will probably require four pairs of Category 5 performance for even higher network speeds.

The wiring pattern for 100BaseT4 is shown in Fig. 4.12. All four pairs are required for this topology, so you cannot use it if you have split your cable into 2 pair jacks. However, it will run fine on any standard TIA 568-A cabling system.

The four pairs of a 100BaseT4 connection are used in a clever manner. The traditional pairs for transmit and receive are maintained on the same pins as the lower speed 10BaseT network, but the other two pairs are used alternately

	Name	Signal purpose	Abbrev
1	Data1+	Dedicated Transmit+	Tx_D1+
2	Data1−	Dedicated Transmit−	Tx_D1−
3	Data2+	Dedicated Receive+	Rx_D2+
4	Data3+	Bidirectional Data3+	BI_D3+
5	Data3−	Bidirectional Data3−	BI_D3−
6	Data2−	Dedicated Receive−	Rx_D2−
7	Data4+	Bidirectional Data4+	BI_D4+
8	Data4−	Bidirectional Data4−	BI_D4−

8-pin Modular jack
(front view)

100BaseT4
Medium-dependent
interface

Figure 4.12 The wiring pattern for 100BaseT4.

for transmit or for receive when sending data. Thus at any one time, three pairs are available for either transmit or receive. This technique is coupled with an advanced line coding to achieve the high bit rate at low symbol rates.

The 100BaseT4 technology suffers from the same distance and repeater limitations as 100BaseTX and has the same speed buffering problem if mixed with slower networks.

Token-Ring

Token-Ring networks were introduced in the mid 1980s by IBM Corporation and others. The network topology is now embodied as the IEEE 802.5 standard. Token-Ring was originally implemented on shielded twisted pair (STP) cable using a unique hermaphroditic connector, commonly called the IBM Data Connector. This cable type and connector is now specified in the TIA 568-A standard. More recently, Token-Ring has been migrated to conventional unshielded twisted pair (UTP) cable (what we simply call twisted pair in most of this book).

Token-Ring has a clever topology that allows an electrically continuous ring to be implemented with wiring that is installed in a hubbed-star configuration. Each arm of the star is called a *lobe*. As shown in Fig. 4.13, a special type of wiring hub, called a MSAU (Multistation Attachment Unit), allows signals from workstations (or from servers or bridges) to be looped through to the next workstation, in turn, until the signal is ultimately looped back to the beginning workstation. This effectively turns the star wiring into a loop or ring, from which the Token-Ring gets the "ring" part of its name.

The "token" part of the name comes from the fact that a so-called token is passed from station to station, along with data, commands, and acknowledgments. The LAN signal thus proceeds in an orderly fashion around the ring in a loop. Here is a very simplified explanation of the process. A station may transmit only when it "has the token." All transmitted frames are passed from station to station around the ring. All stations test each passing frame for messages addressed to them, process the message if it is theirs, and pass a marked token back around the ring. The original transmitting station "releases the token" when it returns around the loop.

The Token-Ring signaling structure is called a Differential Manchester Code, and has no dc voltage component. It can thus be directly, inductively, or capacitively coupled to networking components. This lends itself to the "self-healing" aspect of the physical topology. The MSAU is the key to this architecture. (The MSAU is also commonly called a MAU and pronounced as a word by Token-Ring users, although it is quite different from the Ethernet component by that same name.)

Each port of the MSAU contains a small relay that connects the ring signal to the next port, in turn, when no cable is plugged in. If no cables are plugged in, the MSAU is a small ring within itself. As each cable is plugged into a port on the MSAU, a phantom voltage from the associated workstation opens the

Figure 4.13 A special type of wiring hub, called a MSAU (Multistation Attachment Unit), is used for Token-Ring networks.

relay and the ring signal is diverted down that cable to the workstation. The workstation monitors the ring signal and repeats it back to the MSAU, where it may be diverted to the next active workstation. If a workstation is not active (powered up or inserted), the relay remains closed and bypasses the ring signal to the next port, ignoring the attached cable.

The data rate of Token-Ring is either 4 Mbps or 16 Mbps. A particular ring must operate at just one speed. A ring may be extended to additional MSAUs through the Ring-in/Ring-out ports of each MSAU. Copper repeaters and fiber optic repeaters (or converters) may be used to extend the distances of a ring. Rings are joined together by means of bridges or routers, as with the other network topologies.

The Token-Ring uses two pairs of wires to connect each workstation to the MSAU. The pairs may be incorporated in a shielded or unshielded cable that typically comprises four pairs. The popular IBM Cabling System uses STP cabling in conjunction with special Data Connector jacks and cables to support the traditional Token-Ring installations. The IBM Cabling System is described in Chapter 5. Unshielded twisted pair wiring normally uses an 8-pin, modular (RJ-45 style) plug and jack. This is called the *medium dependent interface connection* in the 802.5 terminology. The wiring patterns of the Data Connector and the 8-pin modular jack are shown in Fig. 4.14.

Station wire for the Token-Ring requires two pairs and the cable may be either STP or UTP. Cables are always run in a star pattern from each workstation outlet to a central wiring closet. The TIA 568-A wiring standard will support Token-Ring networking since it contains four pairs in a compatible pattern. Category 3, 4, or 5 wire will carry either the 4 Mbps or 16 Mbps data rates, although the Category 5 standards offer future growth to 100 Mbps networking. The standard distance limits for station cable in TIA 568-A will take care of Token-Ring networks in most instances, because each workstation acts as a repeater to the next workstation. However, if you wish to calculate the actual working distances according to IBM's recommendations, you must include a complex series of derating factors that take into account the number of wiring closets and MSAUs. These calculations yield the maximum allowable station cable lengths and the allowable cable lengths between wiring closets. Table 4.1 shows a typical planning table for Type 1 STP Cable at 16 Mbps. For a relatively simple network that involves only one or two wiring closets, or if you use repeaters or bridges between closets, you may be able to bypass the calculations. However, if your network is complex, you should consult one of the network planning guides from IBM and other Token-Ring MSAU component vendors.

Because Token-Ring may be run on either 150 ohm STP or 100 ohm UTP, potential impedance-matching problems will exist if you have a mixed-media network. For example, the Token-Ring Network Interface Card (NIC) generally has a 9-pin d-shell connector with a 150 ohm interface. If your station cable (from the wall jack to the wiring closet) is 100 ohm UTP, you will have to use a "media filter" cable from the NIC card to the wall jack to translate from the 150 ohm NIC interface to the 100 ohm cable. Figure 4.15 shows a typical media filter. The media filter actually contains an impedance-transformer (balun) that compensates for the impedance difference. Operating without a media filter can cause serious problems that will limit lobe distances and cause unwanted signal reflections.

Wiring-related problems that may be encountered with Token-Ring networks include excessive distance between workstations and MSAUs or between wiring closets, and problems that occur when an insufficient number of active workstations are on a large ring. Extended distance MSAUs or repeaters may be required to solve these problems. In addition, the normal problems of bad or intermittent connectors, cables, and jacks may affect the network. Token-Ring does exhibit a delay if an improperly wired workstation/cable tries to insert

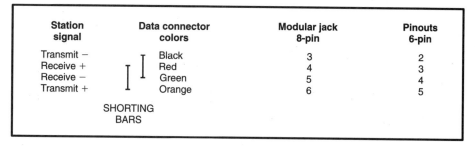

Station signal	Data connector colors	Modular jack 8-pin	Pinouts 6-pin
Transmit −	Black	3	2
Receive +	Red	4	3
Receive −	Green	5	4
Transmit +	Orange	6	5
	SHORTING BARS		

Figure 4.14 The wiring patterns of the Data Connector and the modular jack for Token-Ring networks.

TABLE 4.1 Typical Token Ring Planning Calculation in 16 Mbps Networks. The Maximum Lobe Distances are calculated from the Adjusted Ring Length and the Maximum Transmission Distance (which varies with the numbers of MSAUs and Wiring Closets)

MTD in m (ft) Number of MSAUs	Number of wiring closets				
	2	3	4	5	6
1	130 (430)				
2	130 (420)	125 (410)			
3	125 (410)	120 (400)	120 (390)		
4	120 (400)	120 (390)	115 (380)	115 (380)	
5	120 (390)	115 (380)	110 (370)	110 (360)	110 (360)
6	115 (380)	110 (370)	110 (360)	105 (340)	105 (340)
7	110 (360)	110 (360)	105 (350)	105 (340)	105 (340)
8	105 (350)	105 (350)	105 (340)	100 (330)	100 (330)

NOTE: The maximum lobe distance is calculated for Type 1 and 2 cable as follows:
  Lmax = MTD – ARL
where MTD is the maximum transmission distance from the table.
    ARL is the adjusted ring length (the total cable length connecting wiring closets on the same ring, minus the shortest closet-to-closet length and excluding all patch cords between MSAUs in the same closet).
For Type 3 (unshielded twisted pair) cable, divide the Type 1 Lmax distance by a factor of 2. This factor does not apply to Category 3 and above UTP, which has better transmission characteristics than Type 3/Category 2 cable. Category 3 lobes should observe the 90 m limit of TIA-568A or the recommendations of the MSAU and network adapter manufacturers, whichever is less.

SOURCE: Andrew Corporation

itself into the ring. After a period of time, the offending workstation adapter should deactivate, thus restoring the ring. Some nonpowered MSAUs may get a relay stuck in the open position when no cable is connected. The only way to cure this is to connect a cable from an active station or use a battery-powered test plug to reset the relay.

Arcnet

The Arcnet network type actually comprises a variety of physical wiring topologies that share a common signaling protocol. Arcnet protocol is implemented as a token-passing RF bus that combines some of the features of other network methods, although it was independently developed by Datapoint Corporation in the mid 1970s. Arcnet can be implemented on coaxial cable, twisted pair cable, fiber optic cable, or a combination of the three. The defining factor is the particular type of Network Adapter Cards (NIC) you choose.

Arcnet operates at 2.5 Mbps, although a newer ArcnetPlus at 20 Mbps has been introduced. There are also proprietary implementations at 100 Mbps. The method of protocol operation is similar to Token-Ring, but without a physical ring. All stations are connected to a common bus. When any station transmits, it is simultaneously (more or less) received by all other stations. A token is passed to each station, in a logical sequence, and a station must have the token to transmit. When a transmission is made, the receiving station makes an acknowledgment reply.

The topology of Arcnet can be a star, a bus, or a daisy-chain bus. The star topology may be used with coax, fiber, or twisted pair. The center of the star is either a passive hub or an active hub. An passive hub simply combines all signal inputs (so they may be shared) and provides impedance matching (to minimize reflections). Some of the signal is lost in this process, depending on the number of hub connections, so passive hubs are limited to a small number of ports. Active hubs amplify each incoming signal and distribute it to the other ports. Passive hubs are used with coax cable. Active hubs come in varieties for coax, twisted pair, or fiber. An example of an Arcnet star network is shown in Fig. 4.16.

The Arcnet coaxial bus topology is shown in Fig. 4.17. It is quite similar in structure to Ethernet thinnet, using coax cables with BNC-T's and terminators. However, because it does not provide for any active signal amplification, it is generally limited to about eight stations.

Several types of coax cable may be used for Arcnet, depending upon the particular NIC cards you are using. Arcnet may be used with RG-62/U coax at distances up to 2000 feet, which might be useful as a method to upgrade from IBM 3270 terminals to PC networking without extensive rewiring. Arcnet is also used with 75ohm RG-59/U and the larger RG-11/U coax. A table of cabling alternatives and some of their properties is shown in Table 4.2.

A popular alternative to Arcnet coax is to use twisted pair wire for the transmission medium. An Arcnet twisted pair network can be wired as an active star

Figure 4.15 A typical Token-Ring media filter.

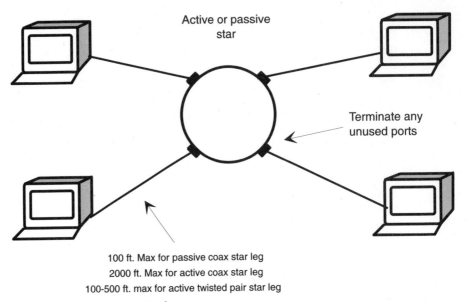

Figure 4.16 The Arcnet star topology.

or as a daisy-chain bus. Technically, a passive star could be used for a small number of workstations, although the active star is more common. A twisted pair star is wired in the same fashion as with other networks. Workstation cables are run outward from a wiring closet or other central location to each workstation outlet. Each cable is terminated at the workstation with a 100 ohm resistor plug. The active star hub includes built-in terminators.

The workstation outlets are sometimes 6-pin RJ-11 jacks with matching cables, since only two pins are used (see Fig. 4.18). Only two pins are required since both transmit and receive occur on the same pair. This is possible, of course, because an Arcnet station may transmit only when it has the token. If you should use telephone flat cable (not really recommended), be careful to get straight through cords, rather than the "flipped" cords that are standard for telephone use. A flipped cord will reverse the polarity of the Arcnet wires and cause the connection to fail.

Standard TIA 568-A wiring may be used with the Arcnet twisted pair star (or with the daisy-chain bus described below). However, most Arcnet twisted pair active hubs and workstation NIC cards use the 6-pin RJ-11 style connector and will require an adapter. An adapter cable may be built for the workstation outlet and the patch panel with a 6-pin modular plug at one end and an 8-pin modular plug at the other.

User and patch cords should be twisted pair wire, as should the station cable. Short lengths of flat telephone-type cord may work for the lower Arcnet speeds, though, since only one pair is used and the problem of crosstalk does not exist.

TABLE 4.2 Arcnet Cabling Alternatives

Cable type	Nominal impedance	Typical connector
RG-11/U Coax	75 ohms	Type N or BNC
RG-59/U	75 ohms	BNC
RG-62 Coax	93 ohms	BNC
IBM Type 1	150 ohms	IBM Data Connector Adapter/Balun
IBM Type 3	100 ohms	IBM Data Connector Adapter to 6-or 8-pin modular plug (RJ-11 or RJ-45 style)
Category 2 (or higher) Twisted pair –2 pairs used	100 ohms	6-or 8-pin modular plug/jack (RJ-11 or RJ-45 style)

The daisy-chain twisted pair bus is a variation of the coax bus using twisted pair. It is useful for small networks with existing cable and in situations where the workstations are all in one room. The one-room design is the easiest to visualize. As with the coax bus, a twisted pair cable is run from one workstation to the next in line, where another cable is run on to the next workstation, forming a chain of workstations. Most twisted pair Arcnet adapter cards have two common connectors, so this is easily done. A 100 ohm terminator plug must be used in the two workstations at the ends of the chain. Of course, the wire can also be run between workstations in different rooms, in the same daisy-chain fashion.

Things are a little more complex if this is attempted with installed wiring. The loops of the chain must be in the same physical station cable. One way to accomplish this is to use a special adapter cable from the wall plate that is split out to two plugs at the NIC card. The cable has two active pairs, say Pair 1 and Pair 2, that are split to the two plugs, both wired for Pair 1 so they will connect to the NIC card's pins 4 and 5. Then at the punch block in the wiring closet, Pair 1 on the first drop is crossconnected to Pair 2 on the second drop, and so forth through the chain. The first and last workstations in the chain will only use one

Figure 4.17 The Arcnet coaxial bus topology.

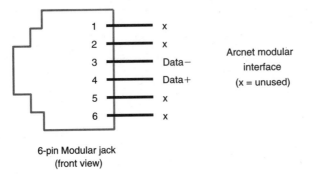

6-pin Modular jack
(front view)

Figure 4.18 The wiring pattern for Arenet 6-pin modular jacks.

of the plugs, the Pair 1 plug for the first and the Pair 2 plug for the last, and will need a100 ohm terminator plugged into the other port on their NIC cards.

Problems in Arcnet cabling are as varied as the many types of cable topologies that are used. The coaxial bus is subject to the same potential problems with BNC-T's and terminators as thinnet. If a terminator is missing, the network will at least partially fail. As with the other LAN types, any break in a bus cable will also cause the network to fail, although the twisted pair bus may be somewhat more tolerant of this. Of course, no signals may pass across the break! If a break occurs, look for damaged cable, connectors, T's, and so on.

Rarely will a bad adapter card cause a total network failure, since the cards are not active until they transmit. An adapter failure will usually affect only that workstation.

Star-type Arcnet networks will usually have a problem only on one leg, although that will take your network down if that leg is the server. Try plugging into another port on the hub. Most twisted pair hubs and some coax active hubs have status lights that will indicate good cable. An active hub can have a total electrical failure, or in some cases have damage to only one port. A passive hub must have the proper value terminator connected to all open ports. Many passive hubs have terminators that are permanently tethered to the hub with a ball chain, so the terminator won't disappear.

Other Network Types

Many other types of networks can operate in the office environment. Some of these other technologies have similarities to standard Ethernet and Token-Ring, but are not truly a part of those networking methods. We will cover a few of the other networks here, including some of the fiber optic topologies. Do not presume that these topologies are out of favor merely by being called "other." In some cases, many technologies, such as Isochronous Ethernet, 100VG-AnyLAN, and ATM, are emerging, and it is simply premature to say which will dominate.

100VG-AnyLAN

A new technology called 100VG-AnyLAN overcomes some of the distance and repeater limitations of 100BaseT and allows operation on 4 pair Category 3 cable. The standard specification for 100VG-AnyLAN is contained in IEEE 802.12.

The 100VG-AnyLAN technology is an interesting combination of asynchronous communication, such as that employed by Ethernet, and round-robin message scheduling, such as that of Token-Ring. The 100VG-AnyLAN topology is an arbitrated star arrangement, that is, a star with transmission control in the hub. Workstations and servers are on the legs of the star, with an intelligent hub at the center. The hub performs the functions of an arbitrated gateway and a switch. A workstation that wants to transmit a message, makes a request to the hub. If the network is idle, the request is granted. The workstation transmits its message and the hub switches it to the proper outgoing port. If more than one station needs to transmit, the hub acknowledges each one in turn, according to a predetermined priority scheme.

The wiring systems supported by 100VG-AnyLAN include 4 pair Category 3 or higher UTP, 2 pair STP-A, and single or multimode optical fiber. The signaling rate on 4 pair UTP is kept low by splitting the signal into four data streams, one per cable pair. The bandwidth requirement of approximately 15 MHz is within the capability of standard Category 3 cabling. The wiring pattern for 100VG-AnyLAN is the same as for any standard TIA 568-A cabling system.

Because 100VG-AnyLAN is not dependent upon a collision detection mechanism, it is not subject to the distance limitations of unswitched Fast Ethernet. Some 100VG-AnyLAN adapter implementations offer dual speed capabilities, as with the 100BaseT products.

Isochronous Ethernet

One of the greatest challenges in integrating voice, data, and even video, is that the technologies heretofore have used entirely different means of encoding and transmitting information. The only practical solution has been to separate those signals in separate cables, each routed to their respective switching or distribution equipment. Isochronous Ethernet proposes to end that divergent situation by combining multiple signals onto the same datastream, so that all three technologies can be simultaneously carried on the same physical medium. Indeed, the technology actually allows quite a number of channels to be carried over the standard 100-meter 4 pair cable link. Isochronous Ethernet is standardized in IEEE 802.9, and the information here reflects the proposed draft revisions to become IEEE 802.9a.

Isochronous Ethernet uses a clever scheme of time-division multiplexing (TDM) and signal encoding to place a full 10 Mbps Ethernet channel plus an additional 6 Mbps or so into the same composite signal. The total data rate in the signal is 16.384 Mbps. A 4B/5B encoding scheme is used instead of the

Manchester coding of standard Ethernet. This yields a 10.24 MHz line transmission rate, thus making the signal well within the capabilities of Category 3 or higher cable.

Two modes of operation are available. In the multiservice mode, the signal is divided into three user channels, the 10 Mbps (actually 10.016) Ethernet P-channel, a 6.144 Mbps bearer C-channel, and a 64 kbps signaling D-channel. The C-channel is easily further divided into 96 channels of 64 kbps subchannels. Of course, the subchannels are exactly the same size as an ISDN bearer channel, so the network can easily support data, digital voice, or compressed video. An alternate isochronous-only mode divides the total signal into 248 bearer subchannels, each of 64 kbps.

A third mode of operation allows the devices to operate as standard 10BaseT, Manchester encoded network interfaces, in full accordance with IEEE 802.3 and with no support for the TDM channels.

Ethernet Fiber 10BaseFx

The increasing use of fiber optics to transport Ethernet signals over extended distances or between buildings has led to the standardization of this method by the IEEE committees. The 802.3 standard now recognizes several variations of the fiber optic link that started out as FOIRL (Fiber Optic Inter-Repeater Link). The replacement for FOIRL is known as 10BaseFL. This standard, as the original name implies, is used to interconnect repeaters, although several manufacturers have used it to link stations as well. Additionally, supplements to the 802.3 standard have added 10BaseFB and 10BaseFP.

The 10BaseFB specification is a "synchronous Ethernet" link between repeaters that extends the limit for repeaters and segments in a single unbridged network. Normal repeaters use a portion of the preamble of the Ethernet packet to distinguish incoming data from noise and to synchronize with the data. Each repeater, in effect, gobbles up some of these "sync" bits, which limits the number of repeaters a packet can traverse. Remember, a repeater does not store and forward a packet, so the lost bits cannot be replaced in the outgoing signal. Synchronizing the links means that the repeater must only distinguish data from noise and the entire preamble may be preserved.

The 10BaseFP specification is a passive star configuration for fiber optics. The signal is shared with other fiber arms of the star via a unique system of optical distribution.

All of these 10BaseF standards use graded-index 62.5/125 micron fiber optic cable with two fibers per link (one for transmit and one for receive). Distance limits are 2 km for 10BaseFL and 10BaseFB, and 500 m for passive star 10BaseFP. The 10BaseFx data rates, of course, are the same 10 Mbps as copper 10BaseT and consequently offer distance and electrical isolation as the primary advantages over copper.

Active point-to-point fiber optic links offer few problems that are not quickly revealed by the link status indicators. The 10BaseFP installation, however, does present the same problems as the coax copper star of Arcnet, since the hub is passive and has no indicators. Fiber optic links offer unique advantages such as extended transmission distances and immunity to interference. However, fiber is not indestructible and may require some special handling during and after installation. For example, fiber optic cable is susceptible to moderate to severe performance degradation from a one-time bending of the fiber below the bending minimum radius. While light may still transfer past the microfracture, transmission loss may soar. Fiber cables may also fracture when they receive a sharp impact. Fiber optic light sources, meters, and even OTDRs (optical time-domain reflectometers) may be required to troubleshoot a fiber optic installation.

AppleTalk/LocalTalk

AppleTalk has turned into a somewhat generic term that is generally used to describe Apple LocalTalk networks. AppleTalk is actually a higher level protocol that can be transported over several types of network topologies. The wiring for classic LocalTalk is a low-speed two-wire bus that is usually run in a daisy-chain fashion from computer to computer. The unique feature of most Apple computers since the introduction of the Macintosh is the built-in LocalTalk networking port. The provision of a built-in networking capability has made printer sharing and file transfer a normal part of operation with Apple computers.

The ease of networking, albeit a slow, peer-to-peer method, has both helped and hurt Apple users. It has helped by making the network a quick and natural solution to data and print sharing that must have made IBM-style PC users envious (or at least confused). On the other hand, the relatively low speed and proprietary nature of the native network, without expensive upgrade modules for Ethernet or Token-Ring, initially put AppleTalk users at a comparative disadvantage.

The original LocalTalk connection used a cumbersome shielded cable, but the Farallon Computing made a compatible adaptation called PhoneNet on twisted pair wiring. This is the simple LocalTalk connection that most users are familiar with. Many users employ standard telephone cord to connect devices. This telephone cord is typically the flat silver-satin type 4-wire cable, although twisted-pair cable will work even better. The network runs at only 230.4 kbps, so the normal crosstalk and attenuation problems are not as severe. LocalTalk can operate well on TIA 568-A wiring.

The same comments apply to potential wiring problems with LocalTalk that were stated for any of the twisted pair networks. Because LocalTalk can use ordinary flat phone wire, many users will simply purchase preassembled cables, which have few faults. However, this wiring method sometimes leads to long runs of wire being routed around furniture and across traffic areas. Any abrasions or cuts in the cable will have network consequences. Often a network must be split at some point and terminated to diagnose such cable problems.

FDDI

FDDI (Fiber Distributed Data Interface) is a fiber optic networking technology that operates at 100 Mbps. It employs dual, counterrotating, token-passing rings. The network concept is actually fairly complex and is different from either Ethernet or Token-Ring.

Network nodes that sit on the primary fiber ring are called *dual-attach nodes* and employ a sophisticated automatic wraparound of the FDDI signal in the event of a cut in the ring. This is somewhat akin to the backup path in Token-Ring, but the fiber in an FDDI ring is expected to take a diverse path to the next node in line. In the event of a fiber cut or other failure, the dual-attach nodes at each end automatically loop back and the network continues without the missing component.

Each dual-attach node has two dual fiber connectors for a total of four individual fiber optic connections. Each connector has a fiber input and output, but the FDDI signal actually passes from one input, through the node processor, and out the opposite connector's output fiber on to the next node. This produces dual signals in opposite directions.

Another type of node, called a *single-attach node*, implements the basic FDDI protocol, but attaches point-to-point in a star fashion to a concentrating node. This concentrating node is customarily on the actual ring (and is dual-attach), and benefits from its fault-tolerant design. The single-attach node lacks the loop back path to its concentrator node, and so is normally used within the same building where fiber cuts would be rare.

The FDDI topology is an ANSI (American National Standards Institute) standard. The fiber is standard 62.5/125 micron fiber terminated in the unique duplex FDDI connector. Both multimode and single-mode fiber are supported, although the interface module is different, since single-mode requires a laser-diode transmitter. An FDDI single-attach connection could be supported on a TIA 568-A fiber optic link, but an adapter cable would have to be used to connect to the 568-SC connectors in the workstation outlet.

Because FDDI is neither an Ethernet nor a Token-Ring protocol, those and other protocols are usually encapsulated into an FDDI packet and decoded at the receiving station. This method simply uses the FDDI link as a transport mechanism that frees one from the point-to-point nature of most WAN links.

ATM

Asynchronous Transmission Mode (ATM) is a very high speed communications protocol that does not really imply a particular LAN topology. As a matter of fact, the ATM protocol can be used over long distances in what is often called the WAN (Wide Area Network). The ATM protocol specifies a 53-byte "cell" that is somewhat analogous to a packet or frame of conventional variable-length MAC-layer protocols. Unlike those protocols, the ATM cell does not always contain source and destination addresses, nor does it contain the higher-level addressing and control of longer packets.

ATM is actually a very abbreviated point-to-point, switched protocol. An ATM message typically sets up a virtual connection between a source and destination, transmits cell-size chunks of the message, and ends the connection. The cells may go through many switching points before reaching the destination. The protocol allows cells to be received at the destination out of order and reordered before being presented to the receiving application.

ATM can be transported at many rates and over a wide variety of media. ATM protocol is implemented over copper and fiber at rates from about 25 Mbps to over 2 Gbps (gigabits-per-second). Because the protocol is not dependent upon a particular connection scheme, many ATM-based technologies have emerged under the ATM umbrella. They are not the same, and should be treated as separate topologies that share a common link basis.

Fiber implementations of ATM often run at Sonet physical parameters, such as 155 Mbps OC3 and 2.4 Gbps OC48. Copper implementations have been introduced at 25, 52, 100, and 155 Mbps. The 25 and 52 Mbps speeds are intended to operate over Category 3 twisted pair cabling. The current ATM UNI-155 standard requires a full Category 5 link for operation. You should be aware that the ATM speeds shown are rounded off from the actual data rates. For example, UNI-155 and OC3 actually run at 155.52 Mbps. To get the exact data rate, multiply 51.84 Mbps by the multiple. For example, OC12 would be 51.84 Mbps × 12, or 622.08 Mbps.

Not to be outdone by fiber optics, copper-based 622 Mbps technologies are under development at this very moment. Watch your mailbox for details! Copper proponents claim that 1 GHz bandwidths are possible for twisted pair, although it is doubtful that this could be achieved with the present complement of connectors and cabling.

Unlike the standard LAN topologies, ATM is a switched protocol. It is not really suitable for standard bus or shared star (repeater coupled) topologies. It does not imply a collision detection or a token passing mechanism. When used for a LAN, it is switched, with the physical switch being at the hub of a star of workstation legs. Often, LAN protocols such as Ethernet or Token-Ring are transported over ATM links via a method of LAN emulation.

5

Cabling Systems

We will describe several LAN cabling systems in this chapter. These systems define the characteristics of cable, connectors, and general wiring schemes without regard to any of the specific LAN topologies that were described in Chapter 4, and all of them also use some form of twisted pair wire as their basis. The general purpose TIA 568-A cabling system will be described, as will several of the proprietary cabling systems that are in wide use.

Many proprietary cabling plans have been offered over the history of LAN wiring technology. Early on, the direction of LAN cabling was uncertain and several major equipment vendors devised detailed cabling systems that met the needs of their equipment. Any customer who followed their guidelines could be reasonably assured of a LAN cable plant that would operate well with that vendor's products. Of course, use of a particular vendor's products also frequently implied use of a particular LAN topology.

Over time, many LAN technologists have learned that the use of a particular vendor's cabling plan might actually limit the utility of their cable plant for other uses. In some cases, adherence to a cabling system might significantly increase system cost over other methods. However, a few cable plans offered wide connectivity of both voice and data. Out of those plans, a universal cabling system finally emerged, EIA/TIA 568, now revised as TIA 568-A.

It is widely agreed that the flexibility and utility of the standard TIA 568-A system offers the best alternative for new installations. The standard can be used in lieu of the proprietary cabling systems, with little difference in performance. It can easily support a wide variety of LAN topologies and offers growth to 100 Mbps and beyond.

TIA 568-A General Purpose Cabling

The TIA 568-A Commercial Building Telecommunications Wiring Standard describes a generic cabling system that can support many types of LANs, as well as many other telecommunications applications. In this section, we will cover the hardware and wiring issues presented by the standard. And of course, the details of many of the wiring components are covered elsewhere in this book. We will rather loosely group the requirements of TIA 568-A and its predecessor documents, EIA/TIA 568, TSB 36, TSB 40, TSB 53, and SP-2840 under the label "TIA 568-A/UTP." The installed link testing requirements of TSB 67 also provide guidelines for this cabling system.

The TIA 568-A cabling system actually includes four types of cable options: unshielded twisted pair (UTP), shielded twisted pair (STP), multimode 62.5/125 optical fiber, and 50 ohm coax. (The 50 ohm coax is grandfathered for existing installations only.) However, the dominance of twisted pair wiring, particularly the unshielded variety, overshadows the other types and will be our focus as a "cabling system."

The TIA 568-A/UTP cabling combines the low cost of modular wiring connections and twisted pair cable with a system of performance classifications and installation procedures. The use of modular connectors and twisted pair wire for LANs was made popular by the availability of inexpensive, existing components that were originally intended for telephone system use. Unfortunately, users soon learned that the cable needed to have specific characteristics, such as minimum twists per foot, to work properly. The cable needed to meet certain performance levels that varied with the application. A cable vendor's rating system, supplemented by Underwriters Laboratories testing, eventually was incorporated into the TIA 568-A/UTP cabling system. The rating system divides wire and components into numbered categories that are associated with increasing levels of performance needed for LAN operation.

Five categories of TIA 568-A/UTP cabling are now specified, as shown in Table 5.1, although low-performance Categories 1 and 2 are specifically excluded from consideration in the standard. Two of these categories, Category 3 and Category 5, are the dominant ones used in most LAN cable installations today. Category 3 can support 10 and 16 MHz networks, such as traditional Ethernet and Token-Ring, while Category 5 supports 100 MHz networks, as well as the lower speeds. Categories are rated in terms of bandwidth, in MHz, rather than data rate, in Mbps, because data encoding techniques often obscure the actual cable signaling rate. The popular 10 and 16 Mbps networks require matching cable bandwidths of 10 and 16 MHz because of their signal structure, but 100 Mbps network techniques can range from 15 MHz (per pair) to 33 MHz because of encoding. Since Category 3 is rated to only 16 MHz and Category 4 only to 20 MHz, some of these 100 Mbps networks require the higher bandwidth of Category 5. Category 5 is also often mentioned as a requirement for some of the 155 Mbps data rates. Components, such as jacks and patch panels, are also rated by category of performance.

Standard 8-pin (RJ-45 style) modular connectors are used for the TIA 568-A/UTP system. Four-pair, 24 AWG, unshielded twisted pair provides the station cable. All 8 wires are connected to the jack. A system of optional punch-down termination blocks and patch panels completes the connection system.

TABLE 5.1 UTP Wiring Performance Categories

Performance	Wire gauge	Nominal impedance	Transmission	Typical uses
Category 1	18-26	Not specified	Audio	Speaker wire, door bells :-)
Category 2	22-26	Not specified	up to 1.5 MHz	Analog telephone
Category 3	22-24	100 ohm ± 10%	up to 16 MHz	10BaseT, 4/16 Token-Ring
Category 4	22-24	100 ohm ± 10%	up to 20 MHz	10BaseT, 4/16 Token-Ring
Category 5	22-24	100 ohm ± 10%	up to 100 MHz	100BaseTX, ATM

Wiring connection is straight-through, which means simply that Pin 1 at one end of a cable will correspond to Pin 1 at the other end. This is true whether we are talking about a user cord, a patch cord, or the in-wall station cable. This is a departure from normal 6-wire telephone cords, which reverse the connections from end to end.

The TIA 568-A/UTP system allows two different color code wiring patterns for the specified 8-pin modular connectors. Figure 5.1 shows the two patterns, called T568A and T568B (not to be confused with the "A" revision of the original standard). The two wiring patterns are electrically equivalent, but the positions of Pair 2 and Pair 3 are reversed. Practically, this means that the color-to-pin correspondence varies, and it makes little difference which you use, as long as both ends of the cable are pinned the same. In other words, if you use the T568A pattern at the workstation outlet (called the telecommunications outlet in the standard), you must also use a T568A patch panel or fanout cable in the wiring closet (called a telecommunications closet in the standard). Pairs 2 and 3 just happen to be the two pairs used by 10BaseT Ethernet, so ignoring this wiring convention will cause the connection to fail. Token-Ring and 4-wire communications circuits, including T1, will also have a failure. Many prewired components are not marked as to wiring pattern, particularly nonrated octopus fanout cables, so be aware what you install.

The TIA 568-A standard defines many wiring infrastructures that apply to all telecommunications wiring, including traditional telephone system wiring. The heirarchy of these wiring structures is shown in Chapter 11. Although the standard creates a very serviceable telephone wiring system, the real need lies in LAN applications which are much pickier about transmission characteristics. The portions of the standard that apply particularly to LAN wiring restrict such things as the length of cable and cross-connects, the characteristics of the cable, and installation techniques.

If you use the TIA 568-A/UTP system for a LAN, you may need to restrict the wiring to a particular category, such as Category 3 or 5. This will place additional requirements on your wiring, such as the maximum cable run, the type of cable, the amount of untwist, and number of crossconnect (punchdown) points that are acceptable.

Figure 5.1 The T568A and T568B wiring patterns.

Figure 5.2 shows a summary of the characteristics for TIA 568-A/UTP cabling. Standard wiring patterns for the plugs, jacks, punchdown blocks, and patch panels are used for the TIA 568-A system. Refer to Chapters 7, 8, and 9 for these details.

AT&T Systemax PDS

The AT&T Premises Distribution System (PDS) is a wiring system that bears great similarity to the TIA 568-A/UTP system. The original PDS system, however, is basically a multipurpose telecommunications wiring system that can also be used for LANs. In contrast, the 568/UTP system is primarily designed with LANs in mind, but can also be used for telecommunications.

PDS is a comprehensive system that covers telecommunications wiring from the building cable entry, through intermediate cable distribution or wire centers, to the telecommunications outlet. It also encompasses campus distribution. PDS uses either unshielded twisted pair wire or fiber optic cable. Copper is used for horizontal wiring to the workstations and fiber is used for vertical wiring between wiring closets. The copper wire is 4 pair, 24 AWG cable, while the fiber is 62.5/125 micron cable. Eight pin modular connectors are used for the copper, and ST connectors are used for the fiber.

The basic layout of a telephone-style wiring plan assumes that there will be hierarchical crossconnect points where large pair-count cables are terminated and fanned out into smaller cables or to station cables. The length of individual runs and the number of crossconnect points are not critical because traditional telephone equipment can run much longer distances than LAN cable.

PDS uses the 110-block as the basic crossconnect component. Station outlet modules use insulation-displacement connectors that snap into a modular faceplate. All connectors are normally specified as 8-pin RJ-45 style connectors, although 6-pin RJ-11/12s are also available. The connector modules are available with markings for "Data" or "Voice" and the rears of the modules are color coded to the wires of the 4 pair cable. The AT&T color code and pin mapping are shown in Table 5.2. Note that the standard wiring pattern for the PDS 8-pin modular jack is identical to the T568B pattern of TIA 568-A. This is not a coincidence, as the influence of standard telecommunications wiring plans, such as PDS, carried great influence in the design of the EIA/TIA standards. An earlier Western Electric plan, referred to as Distributed Inside Wire (D.I.W.), was a logical precursor to both PDS and the EIA/TIA systems.

Figure 5.2 A summary of the characteristics for TIA 568-A/UTP cabling.

TABLE 5.2 AT&T Systimax Structured Cabling System (SCS) Wiring Scheme

Modular twisted-pair cord sets:

4-wire cord		6-wire cord		8-wire cord	
Plug pin	Wire color	Plug pin	Wire color	Plug pin	Wire color
1		1	White-blue	1	White-orange
2	Black	2	White-green	2	Orange-white
3	Red	3	White-orange	3	White-green
4	Green	4	Orange-white	4	Blue-white
5	Yellow	5	Green-white	5	White-blue
6		6	Blue-white	6	Green-white
				7	White-brown
				8	Brown-white

Outlet jacks:

102A-style information outlet (110-type cable terminations), using T568 wiring:

Modular jack pin	Internal lead color	110 Block position	Horizontal wire color
5	White-blue	1	White-blue
4	Blue	2	Blue-white
1	White	3	White-orange
2	Yellow	4	Orange-white
3	Red	5	White-green
6	Black	6	Green-white
7	White-brown	7	White-brown
8	Brown	8	Brown-white

M11BH-style information outlet (H-pattern IDC-type cable terminations), using T568B wiring:

Modular jack pin	Outlet position	Horizontal wire color
1	1	White-orange
2	2	Orange-white
3	3	White-green
4	4	Blue-white
5	5	White-blue
6	6	Green-white
7	7	White-brown
8	8	Brown-white

SOURCE: AT&T

IBM Cabling System

In the 1980s, IBM Corporation designed a multiuse cabling system to support the rapid introduction of twisted pair wiring for computer networks. The cabling system uses a unique hermaphroditic connector (called the Data Connector) and several types of cable identified by a simple numbering system. Figure 5.3 shows a picture of a Data Connector, and Table 5.3 lists the cable types and their characteristics. The cable types, designated Type 1 through Type 9, include shielded twisted pair (STP) and unshielded twisted pair (UTP) cables as well as combinations of both.

The concept behind the cabling system is that most workstation locations actually need at least one data circuit and one telephone circuit. Both needs could be provided by one cable, such as a Type 2 cable, with shielded pairs for the data and unshielded pairs for the telephone.

The cabling system uses a clever genderless connector that eliminates the need for two complementary connector styles (such as the plug and jack of modular wiring). Data Connectors simply plug together to make a connection. This requires that the identical style of connectors be used at the cable end,

Figure 5.3 The Data Connector. (*Courtesy of Hubbell Premise Wiring, Inc.*)

TABLE 5.3 IBM Cable System Cable Types

3IBM type	Pairs	AWG	Impedance	Shielding*	Rating
Type 1	2	22	150 ohms	e/f + o/b	Plenum
Type 2	2	22	150 ohms	e/f + o/b	Plenum
	4	22	600 ohms**	no	Plenum
Type 3	4	24	150 ohms	no	Plenum or nonplenum
Type 4	not used				
Type 5	2 fibers	100/140µ	850 nm	no	Plenum
Type 6	2	26	105 ohms	e/f + o/b	Nonplenum
Type 7	not used				
Type 8	2-flat	23	150 ohms	e/f + o/b	Nonplenum
Type 9	2	26	105 ohms	e/f + o/b	Plenum

*e/f = foil shield, each pair; o/b = overall braid shield
**impedance at 1 kHz, rather than 0.250 to 2.3 MHz.

NOTE: A-suffix cables are rated at higher frequencies than original 20 MHz rating cables. (e.g., Type 1A is tested to 300 MHz.)

wall plate, and patch panel. The unique Data Connector was probably designed with Token-Ring in mind, since that has long been IBM's network of choice. It contains a mechanical shorting-bar mechanism that can electrically loop back a cable or jack when disconnected, an operation very similar to that of the Token-Ring MSAU port.

IBM cabling system installations consist of a series of interconnected wiring closets with star wiring to each workstation location. As we mentioned, either STP or UTP pairs or a combination are contained in each station cable. The STP station pairs are terminated in a Data Connector at each end. The Data Connectors have no gender such as the male/female or jack/plug arrangement of most cabling systems. These Data Connectors are mounted in a flush-mount plate at the workstation outlet and in a patch panel (called a Distribution Panel) at the wiring closet, but are otherwise identical in function to those used on cable ends. The Distribution Panel typically contains 64 mounting positions for Data Connectors that terminate the end of the station cables. It is customarily mounted on freestanding 19-inch rails. The Distribution Panel also contains strain reliefs for the station cables. Data Connector style cables, made using Type 6 cable, are used for patch cords in the wiring closet.

The connection at the workstation is a cable with a Data Connector at one end and, at the other end, an appropriate connector for the particular equipment to be connected. By simply terminating the cable with an appropriate connector, the IBM cabling system can actually support quite a variety of data communications equipment, including coax, twinax, RS-232, and Token-Ring or Ethernet. In the case of Token-Ring, a 9-pin d-shell connector is the appropriate connector at the workstation. The workstation's Token-Ring interface is 150ohms, as is the STP station cable, so no matching balun is needed. At the wiring closet end, many MSAUs are equipped with Data Connector ports and use an ordinary Data Connector patch cable.

At the workstation location, the telecommunications outlet is composed of a faceplate and a regular Data Connector, which terminates the station wire. Type 1 wire terminates in a single Data Connector, while Type 2 wire terminates in one Data Connector and one telephone jack connector. Both connectors are designed to clip into the faceplate. The IBM system offers two basic styles of faceplates. A Type 1 Faceplate has a single Data Connector position, while Type 2 also includes a position for an 8-pin RJ-45 style modular connector for telephone use. Conveniently, the Type 1 outlet is used with Type 1 cable, and likewise, the Type 2 outlet with Type 2 cable. Both faceplates are available in surface mount as well as flush mount options. Figure 5.4 shows a typical workstation outlet for use with Type 2 cable. Other vendors offer Data Connector outlets in several other connector combinations.

Figure 5.4 A typical workstation outlet for Type 2 cable. (*Courtesy of Hubbell Premise Wiring, Inc.*)

The cable and connectors have a matching color code to make wiring simple. However, the cable has a thick insulation and shielding that complicates the termination process in comparison to other types of twisted pair wire. This is undoubtedly why even the station cables are often terminated directly into a Data Connector, rather than a punchdown arrangement as is conventional twisted pair wire. Cable is available in both plenum and nonplenum ratings and must be placed appropriately. Termination of the Data Connector is described in Chapter 10.

The cabling system also includes Type 5 fiber optic cable. The fiber and connectors are basically the same as with other wiring methods. Termination and descriptions of fiber optic cable are described elsewhere in this book.

Several other variations of IBM cabling system wire may be found, the most common being Type 3, Type 6, and Type 9 cable. Type 6 and Type 9 are 26 gauge nonplenum and plenum versions of Type 1 cable. Type 3 wire is 105 ohm, 24 gauge, unshielded, twisted 4 pair cable that meets IBM's specifications. It is available in plenum and nonplenum insulation and is similar but not identical to Category 3 cable. When Type 3 wire is used as station cable, a special adapter cable (called a *media filter*) must be used to match the 105 ohm cable to 150 ohm device interfaces, such as Token-Ring interfaces. Some newer Token-Ring interfaces also have a 105 ohm modular jack, which eliminates the need for the media filter. Similar media filters may be used with 100 ohm Category 3, 4, 5 cable. Note: Older media filters may support only 4 Mbps Token-Ring and will fail when used with 16 Mbps rings.

The IBM cabling system provides some important advantages in a mixed application environment that may use coax, twinax, and twisted pair terminations. The cable has been designed to support all these types of signals using the proper adapters or impedance matching baluns.

However, in a LAN-only environment, such as Token-Ring, the functions may all be satisfactorily accomplished by the TIA 568-A/UTP system at a lower cost. In fact, most estimates place the cost of an IBM cabling system installation at two to three times that of a standard TIA 568-A/UTP installation.

Other Standard Variations

Many other vendors have developed proprietary wiring patterns for cabling of data and voice. Most of these plans support specific features of that manufacturer's equipment and are not general purpose at all. Others incorporate variations of the cabling systems previously mentioned. Of these, two other systems are widely used and deserve mention.

DECconnect

Created by Digital Equipment Corporation, the DECconnect system supports four communication technologies in an attempt to provide a global wiring system. The four technologies are voice, video, low-speed data, and network. DEC's approach uses five different types of cable that are run individually to the DEC faceplate at the communications outlet. The five cable types are listed in Table 5.4, along with the preferred connector.

The complete DECconnect system is rarely used because of its separable cable technologies and its unusual data connector. If you must run a different cable type to each workstation, you might as well keep each technology totally separate. The only advantage the DEC system offers to a mixed media user is a common faceplate mounting for the connectors. The system uses a unique MMJ connector with an offset locking tab and keyway, shown in Fig. 5.5. Unlike a normal keyed jack, with an extra keyway at the side of the opening, the MMJ will not accept a non-MMJ plug at all. It is totally incompatible with all other modular devices, including connectors, jacks, adapters, fanout cables, patch panels, and test equipment. Special MMJ devices must be used, or an adapter cable must be fashioned. (The first M may stand for "maddening.")

Because of the shortcomings of the DECconnect system, DECconnect wiring is primarily used to connect DEC asynchronous computer terminals using a variant of RS-232/RS-422 wiring. These async terminals use an MMJ connector for the interface. The DECconnect system specifies a clever symmetrical wiring pattern for the data connector that avoids the failure that would result from reversing the order of pin wiring, a common error. Ironically, the most frequent source of this type of error would come from the use of standard telephone cables that flip pins from end to end. These are the very cables that will not plug into the MMJ outlets.

NORDX/CDT (formerly part of Northern Telecom), IBDN

Another variant of twisted pair wiring is the Integrated Building Distribution System put forth by Northern Telecom, often referred to as Nortel. The IBDN components are now marketed by NORDX/CDT. This system is quite similar to the AT&T PDS and primarily supports telecommunications wiring. As we saw with PDS, data networks can be run over such systems, but the distance limitations and other installation practices must be observed.

IBDN uses the common system of modular connectors with cross-connects based around the BIX block. IBDN also contains fiber components,

TABLE 5.4 DECconnect Cable Types

DECconnect type	Cable	Connector
Voice	4 pair	RJ-45
Video	75-ohm coax	F-type
Low-speed data	2 pair	MMJ (modified RJ-45-type)
Network	50-ohm coax	BNC
	62.5/125 micron fiber	ST or SMA

Figure 5.5 The DEConnect system uses a unique MMJ connector with an offset locking tab and keyway.

as does the PDS. While the IBDN system is not specifically designed to match TIA 568-A requirements, it will allow common types of network connections if the distance limitations for the network are not exceeded. When used in accordance with the guidelines in the standard, an IBDN installation can be said to be TIA 568-A compliant.

All of the modular connectors, jacks, cable, and color codes of the AT&T PDS system are common for IBDN. The interconnection component, of course, is the BIX block, rather than the 110 block. Please refer to the PDS section in this chapter and the BIX block section of Chapter 8 for more information.

6

LAN Wiring
for the Future

It has been said that the only thing that remains constant is change itself, and that's certainly true in reference to changes in computer technology. Even so, it is possible to accurately predict the near-term future of LAN wiring because it is visible now.

One of the interesting features of technological change in the computer industry is the steady progression from experimental technologies to leading-edge and then to widely accepted. While experimental technologies are quite divergent, those that are feasible eventually become the hot new technologies that define the true state of the art. Of course, a new feature or capability does not immediately supplant existing products, but eventually the item becomes readily available at a reasonable cost. At that point, standardization emerges, whether formal or de facto, and the product becomes widely accepted.

The game, as it were, is really not in trying to guess which experimental developments will become accepted, but rather in observing which leading-edge technologies have begun to be accepted and then employing them in the future. The creation of recognized standards is the key to what technologies will be used in the future.

In local area networking, we are fortunate because we must operate in a multivendor environment. The need for interoperability means that we must have strong, internationally accepted standards for networking. Another fortunate circumstance is that no proprietary technology dominates the field. Many, if not most, of the technologies that have become widely used are a result of unprecedented industry cooperation through partnerships, consortia, and standards working committees. In addition, because of the need for interoperability, standards tend not to change radically once they are implemented. This is particularly true of LAN wiring.

You can expect your LAN wiring system to remain totally usable over many years if your needs remain relatively constant. This is not to say that if you progress from simple printer and file sharing to high-bandwidth graphic or video applications that you might not need to upgrade some network components. However, most local area networks support organizations with relatively constant needs. For example, if you have a standard 10BaseT network in place today, you may reasonably expect it to remain viable for a number of years.

On the other hand, if you are installing new network wiring, you should plan for the future. For LAN wiring, the future is fairly clear. There will be a gradual move from the current 10/16 Mbps network speeds to 100 Mbps and higher speeds. There may be no immediate need (or budget) to convert initially to 100 Mbps, but you should probably not install any new cable that will not serve that speed.

So, how do you plan for the future? How do you put in a 100 Mbps network, or one for 155 Mbps? Should you use copper wire or consider fiber? Can you hedge your bets? Let's see.

100 Mbps Networking Compatibility

If you currently are using 10 Mbps Ethernet or 16 Mbps Token-Ring, you may not be using LAN wiring components that will move you to higher speeds such as 100 Mbps. These lower speeds are much less demanding in their need for transmission media than the high-speed networks, such as Fast Ethernet.

For example, 10BaseT will run at a workstation-to-hub distance of 100 m on Category 3 cable. Some users even report up to 150 m distances as usable, although this exceeds the limit set by standards. On older telephone cable that roughly meets current Category 2 standards, 10BaseT can usually operate 50 m (over 150 ft) with few problems. If you try to upgrade such a network to 10 times the speed, however, it will surely fail.

So, what can you do if you may need to upgrade to 100 Mbps? The answer is to install a LAN wiring system that meets the performance specifications for the higher speed now; it will operate very satisfactorily at the lower speed for the present and will provide additional service as you upgrade to high speeds.

One sure way to provide such a wiring system is to use the TIA 568-A revised standard for a Category 5 cable system. Category 5 wiring is designed to support full 100 MHz/100 Mbps speeds, and is also able to support some applications at 155 Mbps and beyond. At the same time, Category 5 wiring will fully support lower network speeds, such as 10BaseT, Token-Ring, and Arcnet, as well as intermediate speeds for some of the ATM and Fast Token-Ring proposals. For that matter, it will also support voice, ISDN, and T1/E1 circuits very nicely. Table 6.1 shows a table that summarizes the supported uses of Category 3 and 5 installations.

True Category 5 wiring should be used for a single application, not shared between data and voice, for instance. If your needs are uncertain, you may wish to overwire your workstations by pulling more than one cable to each workstation location. There is a school of thought that holds that total separation of voice and data facilities—down to separate wiring closets and equipment

**TABLE 6.1 Common Applications for EIA/TIA 568A
Category 3 and Category 5 Facilities**

Category 3	Category 5
1.5-Mbps IBM 3270*	All Category 3 Applications
2.5-Mbps Arcnet	25 Mbps UNI ATM**
4.0-Mbps Token-Ring*	52 Mbps UNI ATM**
10-Mbps 10BaseT	100BaseTX
16-Mbps Token-Ring*	100BaseT4**
25-Mbps UNI ATM	100 VG AnyLAN**
52-Mbps UNI ATM	155 Mbps UNI ATM
100BaseT4	
100 VG AnyLAN	
Baseband Voice	
ISDN Basic Rate Interface	
ISDN Primary Rate	
T1/E1 Carrier (1.544 Mbps)	
RS-232D (partial interface)	
RS-422	
Baseband Video*	

*Application may require baluns or special interfaces.
**Application may offer better performance on Category 5 facility.

rooms—is the safest course. This certainly might make sense if your telephone system and LAN system are maintained by separate entities. Standard practices for telephone wire installation are much looser than the strict requirements for a Category 5 data installation. Many LAN managers have had bad experiences when the two types of installation methods are mixed. However, if all of your cabling is installed and/or maintained by the same people, you should be able to ensure that Category 5 wiring stays Category 5.

Ensuring the performance to Category 5 standards requires much more than just Category 5 components. Location, routing, and workmanship issues are at least as important. These three issues are difficult to judge because they are related to the installation process. You can buy the finest certified components and still have them installed improperly. That is why we have devoted an entire part of this book to the installation of the wiring devices.

Category 5 testing and certification

The simple proof of 100 Mbps network performance is to connect your 100 Mbps hubs and workstations and see if they work. Unfortunately, many of us must install cabling systems long before we buy the first 100 Mbps hardware. Even if high speed hardware were to be installed immediately, the time to cure cable problems is during the build-out stage of construction, not after move-in.

The solution is to have the cable system tested and certified as part of the installation process. This testing and certification process is described in more detail in Chapter 12, but we can summarize it here.

The first step in certification is to use the proper materials. All cables, jacks, patch panels, patch and user cords, and connecting blocks should be certified to the proper category by the manufacturer. There is no major benefit to specifying all this by brand name unless you are using a particular manufacturer's connecting hardware system that offers unique features. For a smaller installation of 50 station drops or less, this will probably just raise the price and narrow the field of bidders.

The second step is to use properly trained installers, particularly for Category 5 installations. Several certification programs exist for cable installers. The vendor-specific programs generally emphasize a common set of workmanship techniques and are adequate training in most situations. You should not insist that installers be certified by a particular vendor unless you are willing to put up with the extra expense and aggravation of such a policy; it is more important that each installer either be formally trained at a recent course, or be closely supervised during the installation.

A few industry-wide certification programs exist, including the Registered Communications Distribution Designer (RCDD) program from BICSI, Inc. (Building Industry Consulting Service International, Inc.). The RCDD requires experience, successful completion of a test, and membership in BICSI. A LAN Specialization program also exists. Plans are underway to introduce an installer certification program that does not require membership in the organization.

The third step is to individually test and certify each workstation link. The worst-case link performance criteria and testing guidelines of TIA 568-A and TSB-67 are an excellent basis for certification. Each link should pass all the tests specified for the category of the installation and you should get a printed report. Keep in mind that you will have to pay extra to have each link scanned. The equipment to perform the testing is expensive and the installer will expect to be compensated.

155 Mbps Networking

Data rates of 155 Mbps are usually associated with ATM protocols. In fact, the use of the 53-byte ATM protocol is not dependent upon any particular data rate or medium. However, if you want to use one of the ATM-based 155 Mbps connection methods, you will need to make certain that your cable is up to the task.

The best guideline for specifying a 155 Mbps cabling system is to be sure what category of cable is required. Data speeds above 10 or 16 Mbps may not necessarily require Category 5 cable. Several high-speed technologies actually split up the signal onto three or four pairs, rather than the customary two. These methods may operate over Category 3 cable at reasonable distances, even though the apparent data rate is 100 Mbps or more. Both 100BaseT4 and 100VG-AnyLAN are examples of this method.

In addition, special encoding methods may reduce the signaling rate on the cable, while maintaining a high data rate. For example, CAP-64, a 155.52 Mbps proposed ATM Forum standard, uses a cable signal of 30 MHz or less.

In order to support signaling rates of 155 Mbps and beyond, it may eventually be necessary to go beyond the Category 5 specification. However, it is more likely that new, more capable transmission and encoding methods may allow the advanced levels of performance without upgrading the cable and connecting devices. Once Category 5 cable becomes the bulk of installed cabling, the ability to use it at higher speeds will become even more important.

"Enhanced" Category 5 Cable

The Category 5 performance specifications were undoubtedly chosen to meet the minimum needs of 100 MHz bandwidth transmission. Rather than pushing the performance requirements of twisted pair cable to some theoretical maximum level, the Category 5 certification simply requires materials and construction that are adequate for present-day high-speed networks, such as those operating at 100 to 155 Mbps. As we have mentioned in the text, such data rates use special encoding techniques and multiple-pair transmission to get the actual frequency requirements well below the equivalent bits-per-second number. That is why some of the 100 Mbps techniques actually run on Category 3 cable (which is rated to only 16 MHz). Other high-speed data standards do require Category 5 cable, but only because they exceed the 16 MHz Category 3 and the 20 MHz Category 4 specifications, not because they really use all of the 100 MHz Category 5 bandwidth.

Nevertheless, it is quite possible to build better unshielded twisted-pair cable, and offer that enhanced-performance to the market at enhanced prices and enhanced levels of marketing. The enhanced Category 5 cable goes by fancy names that imply "plus" or "extended" performance, or mention a number obviously higher than 100 to show their higher frequency range. Some manufacturers have even gone so far as to refer to their cable as "category 6" (note the small "c"). There is currently no such thing in the TIA standards, but there may be in the future. However, such an extended category specification may use different connectors than the ubiquitous 8-pin modular connector system.

In fact, the most significant limiting factor in higher-frequency link performance over the TIA 568-A/UTP system is actually the connector, not the cable! The connector was never designed to operate at high frequencies, and, with the additional problems caused by the customary pin-to-pair assignments, the connection produces detrimental levels of crosstalk, untwist, and imbalance.

A link may perform better with enhanced Category 5 cable, particularly with regard to attenuation. However, how much better than adequate do you need? The link performance consists of several criteria, and the limiting component is generally the connector, which no amount of cable enhancement can improve. If and when the standards committees decide to expand the categories, it is anticipated that a different connector with much higher performance will be used. Hopefully, the connector will be as easy to use and relatively inexpensive as the current choice.

The Future of Copper Wiring

In a paraphrase of Mark Twain, the report of the demise of copper wire is greatly exaggerated. There has been a great deal of press for several years that copper wire cabling should be abandoned in favor of either fiber optic cabling or wireless networking. While the claims of the proponents are not without merit, they are also not unbiased. The loudest supporters of copper-less networking are the companies that produce those competing technologies. It is only natural that they would see their products as being technologically superior to older methods.

Truth, as often is the case, probably lies between the extremes. There is a certain philosophical attractiveness to alternatives for copper wire. The fiber alternative offers a considerable increase in bandwidth, which translates to an increase in LAN data rates. Unfortunately, it may come at greater cost and more limited flexibility. No one would argue that it would be difficult to adapt the hub, patch panel, and workstation outlet to fiber. Fiber has its own set of installation and testing difficulties, in addition to a higher component expense. Most of your existing networking equipment would have to be scrapped and replaced with optically-interfaced equipment and adapters. Advantages in one area may be disadvantages in another. For example, the voltage isolation inherent in fiber optic cabling would make it impossible to centrally power telephone instruments, so communications could not be maintained in a power outage.

The solution, for the time being, is that fiber has an important place in telecommunication systems, but it will not immediately supplant copper wire as the cable of choice. Fiber will continue to provide interbuilding links, links between far-flung hubs, and links needing greater bandwidth/distance performance than copper can provide.

The wireless alternative is an intriguing one. It eliminates the need for any direct placement of cable, whether copper or fiber. It is heavily promoted as the future of networking. However, it has serious inherent disadvantages. The first disadvantage is that it knows no bounds. That means that your network connection signal can travel far beyond your immediate office (even outside your building). Not only is the signal transmitted where it is not wanted, but the wireless LAN adapter is also susceptible to outside receive interference. Second, as more and more wireless devices are placed in close proximity ("close" can be hundreds of meters in distance), more interfering signal clashes occur. Do you remember the early days of cellular radio, when the service actually met its claims of clear, interference-free communication from almost anywhere? The addition of thousands of users has greatly degraded the quality of reception. The same degradation happens to crowded wireless LANs, except that the interference cannot be heard; it simply slows down the network in silence.

The third disadvantage is troubleshooting and management. A direct cable network is easy to manage and repair. With modern hubs, problems in the network cabling are easily isolated to one station, and monitoring of the process is straightforward. Finding a radio-frequency problem in a random, deterministic, virtual network is a much more challenging task. Fourth, wireless networking is very expensive in relation to direct cabling. Wireless network adapters are usually 5 to 10 times the cost of copper cabling systems, including the adapters.

Wireless networking, however, has a clear advantage in places where running cable is impractical, impossible, or prohibitively expensive. It may be used as a very cost-effective link between buildings, especially across public rights-of-way. Practical or aesthetic reasons may tilt the scales toward wireless. For instance, a connection in a large arena or auditorium might be much easier to accomplish with a wireless link. Finally, if the workstation is mobile, wireless is crucial. (Picture a forklift-mounted workstation with a network cable trailing along behind!)

No, the demise of copper wiring is not imminent, but neither does it meet all possible needs. Each technology should be applied where it is the most appropriate and cost-effective.

Migrating to Higher Speeds

The Queen kept crying 'Faster! Faster!'
... 'Are we nearly there?' Alice managed to pant out at last.
'Nearly there!' the Queen repeated. 'Why, we passed it ten minutes ago!
Faster!'

Higher network speeds are upon us whether we like it or not. Actually, many data-transfer applications have begun to overload some networks, so the change may be welcome indeed. However, few network managers have the luxury of being able to totally upgrade their networks to the higher speeds of 100 Mbps or higher. These networks will have to be gradually migrated to higher data rates, as money and the availability of technology allow.

Migrating to higher speed networks is a much broader subject than the LAN wiring issues being covered in this book. So, for simplicity, we will outline only the wiring implications of high-speed networking. This table may help.

Minimum Category	Pairs	LAN speed	Supported technology
3	2	10 to 25 Mbps	10BaseT, Token-Ring, ATM-25
3	4	100 Mbps	100BaseT4, 100VG-AnyLAN
5	2	100 to 155 Mbps	100BaseTX, ATM UNI-155
5	4	622 Mbps	ATM UNI-622 (proposed)
Fiber/MM	1	10/16 Mbps	10BaseFL/FB/FP or Token-Ring
Fiber/MM	1	100 Mbps	100 Mb Ethernet, 100VG-AnyLAN, or FDDI
Fiber/MM	1	155, 622, Mbps	ATM-155, 622, SONET/OC3,12
Fiber/MM	1	2.3 Gbps	SONET/OC48

Of course, there are lots of variations for many of these networking speeds, and some have not yet reached prime time, but this gives you an overview. How can you use this information to plan a migration to higher speed network cabling? First, unless you are designing a completely new facility, or planning to replace all of your existing network cabling, you should find where on the chart you are right now. For example, if you currently have 4 pair Category 3 cable (real Cat. 3, not assumed Cat. 3), you can upgrade from 10 or 16 Mbps to 100 Mbps only by using one of the 100 Mbps/Category 3 technologies shown. You will not be able to use 100BaseTX or ATM-155 at all, without replacing your cable with new Category 5 cable. On the other hand, if you already have Category 5 cable, you can upgrade to any of the 100 Mbps or 155 Mbps technologies; that is, if that cable was installed to Category 5 standards.

continued

> If you already have fiber installed, you are probably set for any of the currently envisioned higher speeds. At the worst case, you may have to adapt the connectors or reterminate the fibers with newer SC connectors. You can also get to the higher speeds by installing new fiber, or by including a fiber run along with conventional twisted pair to each workstation area.
>
> New installations that will initially employ lower-speed networking have a special problem. How can the network manager plan for the future implementation of these higher speeds when the exact technology choice is not yet known? The best bet is to put in the highest grade facility you can afford. It would be ideal to pull both fiber and copper to each work area. However, if your budget limits you to twisted pair, you should at least put in a full Category 5 cabling system, so you will get all the bandwidth possible. And remember, any decisions you make now may still not keep up with the future, but perhaps they will at least last through the 5-to-15 year useful life projected for most cabling systems.
>
> *'A slow sort of country!' said the Queen. 'Now, here, you see, it takes all the running you can do, to keep in the same place. If you want to get somewhere else, you must run at least twice as fast as that!'*
> *... and Alice began to remember that she was a Pawn, and that it would soon be time for her to move.*
> *Lewis Carroll, Through the Looking Glass*

Future (Wiring) Shock

Alvin Toffler, in his famous work, *Future Shock* spoke of the pace of technological advancement as a shock to the individual and to society. In some ways, we in the computer industry have become accustomed to the rapid obsolescence of our equipment and methods. Computers, printers, and even network hubs are relatively easy to upgrade or replace, as they are movable objects, attached only by their wire connections. The LAN wiring is more of an immovable object, fixed into place in walls, ceilings, and cubicles. What can be the result when the irresistible force of change meets your wiring? How can you prevent this from happening?

Well, one thing you can do is to plan for the future. Design new cable systems so they meet tomorrow's needs as well as today's. If you are installing a 10BaseT or Token-Ring network, that means you should install a cable facility that can go to the next step, 100 Mbps. If you have an existing facility, add Category 5 cabling and hardware when you add workstation drops. Plan the additions so the old cable can eventually be converted to the new system. A graceful migration such as this can be accomplished over a period of time to minimize the budget impact.

Proper planning can give a cable system an extended life. Providing for the future may cost a little more now, but it will simplify change when it eventually comes.

The tubing pull

In some ways, it is impossible to predict every twist and turn technology will make. It would be nice to have some way of replacing the wiring periodically if needed. You might need to upgrade to fiber optic cable. Maybe each workstation will need another cable and outlet to add an application such as video. What if you could simply go to the wiring closet and pull in some new cable?

It may be possible to do just that. A technique that works well in closed-ceiling buildings can also be extended to conventional grid-ceiling buildings. As the building is constructed, continuous plastic tubing runs are made from the wiring closet to a flush-mounted outlet box at each workstation. A pull string may be blown through the tube, or pulled in with a snake. The cable is then pulled through the tubing to the outlet and terminated normally on both ends. As needs change, new cables are pulled through the tube using the old cable as a pull string.

This method works very well in closed construction, which is very expensive to rewire. It also works in smaller offices where the cable runs are not too long. It may have a special advantage for Category 5 runs, as it avoids the performance-robbing problems of kinks, minimum bend radius, tight tie-wraps, and metallic runs along pipes and structural steel.

Wiring Devices

Station Outlets

In this chapter, we begin to go through the chain of wiring devices that make up a LAN wiring connection. The first item in our list is the station outlet placed at the workstation area. The TIA 568-A standard requires that each work area be provided with two outlet jacks. One jack must be a 4 pair UTP connection. The other jack may be another 4 pair 100 ohm UTP connection, or it may be either a 2 pair 150 ohm STP-A connection, or a 62.5/125 micron fiber optic connection.

The station outlet is referred to by several names. The TIA 568-A standard refers to the outlet as a telecommunications outlet/connector, to emphasize that it must include the actual connector jack itself. We refer to the outlet in this book by the name *workstation outlet* or simply *station outlet*. This usage implies that the connector jack is not directly attached to the workstation or other network device. Such an attachment is made by an additional cable called a *user cord* or *equipment cord*. The modular or other jack typically attaches to a mounting plate or faceplate, which mounts in an enclosure. The entire assembly, the jack, plate, and enclosure, forms the station outlet.

We will cover the mountings and jack types first. We will also discuss mounting methods for the jack/faceplate assemblies, including surface and flush mountings. Finally, we will detail outlet identification and workmanship issues.

Jack Mountings

The connector jack at the station outlet is mounted in several ways. We will cover the mounting first because it is useful to have an understanding of connector-mounting methods when we actually begin to discuss the jack types in the next section. In this discussion of jack mountings, we are actually talking about the faceplate/jack assembly in those instances where a separate faceplate is used.

Flush mounts

Flush-mounted jacks are the familiar wall-mounted jacks we often associate with commercial telecommunications wiring. This type of jack arrangement places the jack faceplate flush (or even) with the wall. Typically, the jack extends into the wall and mounts via machine screws to an underlying outlet box or mounting ring.

Mounting rings and outlet boxes

The mounting ring or outlet box provides a secure location for mounting the station outlet. Several types of each are available. Generally, either type of mounting is acceptable. Telecommunications wiring is a low-voltage, low-amperage wiring that is not required to be routed in conduit or mounted in an enclosure by the NEC. However, your local construction codes may require that the outlet be mounted in an enclosed box in some cases.

Traditional outlet boxes are the same design as those used for electrical circuit wiring. The boxes are available in metallic or plastic styles and different designs are available for prewiring, during initial construction, and for postwiring, when the building is complete. Figure 7.1 shows several types of outlet boxes that are suitable for telecommunications use. The 2-inch wide box is the most common for this style of wiring, although a 4-inch square box with a 2-inch wide mounting ring may also be used. In general,

Figure 7.1 Several types of outlet boxes that are suitable for telecommunications use. *(Courtesy of The Siemon Company.)*

the telecommunications outlet plates that mount to these boxes are designed for the 2-inch opening. The opening sizes are also referred to as 1-gang (2-inch wide) and 2-gang (4-inch wide) in the building trades.

Outlet boxes for prewire use will have nail plates for attachment to vertical supports or studs in the walls. Some of these boxes have captive nails already attached. The box should be securely attached to its support and positioned to allow for clearance through the wall sheathing. After the wall is complete, the mounting plates of the box should be even with the wall surface. The boxes should be positioned at the same height as required for electrical outlets. This might not necessarily be a construction code requirement, but it enhances the appearance of the final installation. The cable may be run through a bare knock-out hole in the box or through a conduit or flexible cable connector. Plastic sleeves are also available to minimize any chance that the cable might be damaged by the sharp hole edges.

Outlet boxes for postwiring use are also available. These boxes are sometimes called *old work* or *sheetrock* boxes. Because they mount in hollow walls that have already been constructed, they must be inserted from the front of the wall. An opening is cut in the wall big enough to insert the box body. Additional room may be required for the mounting device that secures the box to the wall board. These mounting devices are usually screw-tightened to clamp the mounting flange of the box to the wall.

Another type of flush mounting device is the mounting ring. Mounting rings are available in several varieties, as shown in Fig. 7.2. Standard mounting rings provide a target location for telecommunications outlets during construction. These rings have mounting plates that allow them to be attached to the wall's vertical supports. The mounting ring normally has a two-inch by four-inch opening that is formed into a flange. The flange protrudes forward from the wall support and is the right depth to be flush with the front of the wall surface, when it is later attached. The top and bottom edges of the ring have mounting holes to which the mounting screws of the outlet attach.

Mounting rings for postwiring are also available. These rings have flexible metal fingers at the top and bottom of the opening that are part of the stamping. A rectangular hole of an appropriate size is cut into the wall board. Then, the mounting ring is placed in the hole until it is flush with the wall. The metal fingers are bent around into the hole and back against the rear surface of the wall board to secure the ring. An alternate design has a vertical screw slot incorporated into the metal fingers. After the fingers are bent into the mounting hole, a provided screw clamps the front flange of the ring tightly to the wall, using the slotted finger as a rear clamp. This prevents any undesired movement in the ring that would occur with the other type of ring.

Prewire and postwire methods

Prewiring may be accomplished with actual cable runs or with a combination of pull strings and late-stage cable runs. If the actual cable is used for the prewire, it must be installed at the stage between wall support installation

Figure 7.2   Mounting rings are available in several varieties. *(Courtesy of Erico, Inc.)*

Figure 7.2 *(continued)*

and wall sheathing attachment. This is often referred to as the *rough-in stage*. With the wall supports in place, the cable is run from each wiring closet to each served station outlet box or ring and either coiled inside the box or secured to the ring. After the wall sheathing is in place, and usually after the wall has been finish painted or papered, each cable is trimmed to working length and terminated in the jack. The assembled outlet is then mounted in the box or ring.

Using the actual cable in the rough-in stage can present some problems if you are planning a full Category 5 installation. The cable is left exposed and subject to damage by others during the construction. In addition, you must coil up the wire very carefully to avoid the sharp bends and kinks that can disturb cable performance at this category of operation.

Another technique is to leave a pull string in the wall during the rough-in stage—placed in a hollow wall or inside a riser that connects to the outlet box. Outlet boxes may have vertical risers (conduit) that run from the top of the box, through the wall's top-plate, and half a foot or so into the plenum space above the ceiling grid. A pull string may be run through the conduit to assist in cable installation. If a mounting ring is used for the outlet, a pull string is particularly convenient. The opposite end of the string should be secured to anything above the top-plate, where it will be visible to the cable installer. After the wall construction is finished, and preferably before the ceiling tiles are in place, the cables are run from the wiring closet to each outlet's pull string, attached, and pulled into the outlet where they may be terminated.

Pull strings and conduit are effective ways to run cable into difficult outlet locations, such as below windows or in columns. Cable may also be coiled and secured above a planned outlet location for later pulling into an outlet box or ring. Again, if you are planning a Category 5 installation, be careful not to put sharp bends or kinks into the cable. Tie wraps may be used sparingly, but should not be overly tightened.

You should take care when running your cable into an outlet box so that you do not slice through the cable jacket. Metallic boxes often have sharp edges as a result of the way they are manufactured. These sharp places can cut into the cable and short some of the wires. Mounting rings may present other hazards to cable, such as the sharp screws that are used to secure some of the rings to the wall sheathing.

Surface mounts

What do you do when you cannot mount the outlet jack inside a hollow wall? You can use an outlet box that is specially designed for mounting on the surface of a wall. These mountings are called *surface mount* boxes and come in two varieties, self-contained and surface adapter boxes.

The self-contained surface-mount outlet contains one or more connector jacks for the type of telecommunications or LAN service you are installing. These boxes are often called *biscuit blocks* because of their appearance. A typical surface box, shown in Fig. 7.3, contains two modular jacks, one that may be used for data and one for phone. The TIA 568-A standard calls for two circuit links to each workstation, but leaves the use of the links up to the user. Both should be fully wired, 8-position modular jacks to satisfy the standard, although you may wish to make one a 6-position jack if your telephone system calls for that. This is particularly appropriate if you choose to totally separate the telephone and data wiring, as we discussed earlier.

Self-contained outlet boxes are appropriate for solid walls, modular furniture, and other difficult to reach areas. Care should be taken to protect the wire with a surface raceway if it is exposed. Providing strain relief to the cable is a significant problem with these boxes. Standard practice is to tightly secure the cable at the point of entry to the box, with a tie-wrap or a knot in the cable. This would obviously distort the cable jacket or break the minimum bend radius of the cable allowed by the standard. If the box can be secured to a surface, a better approach might be to use a surface raceway to protect the cable up to the point of entry to the box and leave the cable entry relatively loose. These surface raceways are described below.

The other variety of surface-mount box, shown in Fig. 7.1, is a rectangular plastic box with a 2-inch by 4-inch opening for a standard, flush-mount outlet plate. These surface adapter boxes are really part of an elaborate surface

Figure 7.3 A typical surface box containing two modular jacks, one that may be used for data and one for phone. *(Courtesy of Leviton Telcom.)*

raceway system that is technically suitable for electrical wiring in addition to telecommunications wiring. The system includes long, flat plastic raceways that may be joined together with a variety of couplings, including corners and straight couplers. Surface raceways are available from several manufacturers. An older system of metal raceways and boxes exists, but the plastic ones are much easier to use and are permitted in almost all jurisdictions for Class 2 or 3 wire or telecommunications wiring. Many of the plastic raceways have an adhesive foam tape backing for quick and easy mounting to the wall or other surface along which they run. The boxes may also come with foam mountings, but you are better advised to secure them to the wall with screws. The screws also help stabilize the entire raceway system.

Other plastic raceways are available that do not interlock with a surface box. These raceways use the same type of foam-tape mounting, but snap closed after installation so that the wire(s) can be slipped in at the side of the raceway after it is stuck to the wall. These raceways are more appropriate for the self-contained surface outlets.

A note of caution should be stated regarding these raceways and matching boxes. The safety standards of the NEC and other agencies do not allow telecommunications wiring to share any raceway space or outlet box with Class 1 wiring. Class 1 wiring is wiring for lighting or electrical power circuits, the typical line-power that supplies lights and electrical outlets. In some narrow cases, the wiring may be allowed in the same outer raceway cover, but it must be separated by a specific type of partition. See the code books for a detailed explanation of when and where this can be done. Bear in mind that the run-of-the-mill square power pole that drops from a suspended ceiling to a work area rarely meets this test, even though there is plenty of extra room inside for the telecommunications cable. Some modular furniture may have the same problem, as the wiring channel may be occupied by Class 1 wiring. In such cases, you may have to run the telecommunications wiring external to the wiring channel. Several varieties of multimedia (copper and fiber) outlet boxes are shown in Fig. 7.4.

Modular Jacks for Twisted Pair

Modular outlet jacks for twisted pair wiring have been around since modular connectors were first introduced several decades ago. The first of these jacks had loose wires from the connector to screw terminals mounted to the body of the connector assembly. Because these jacks were for analog telephone use, at very low frequencies, little consideration was given to the crosstalk and impedance issues that are very important at today's higher frequencies.

Jacks for LAN use are very different from these old telephone styles. In this section, we will describe the various types of jacks that are used for twisted pair LAN wiring. For our purposes here, the jack will include the modular connector itself, the means of terminating a twisted pair cable, and the wiring in between.

Figure 7.4 Examples of flush mounted and surface mounted multimedia (copper and fiber) outlet boxes.

Jack plate types

There are three basic types of mounting plates for modular jacks: fixed plates, modular plates, and modular assemblies. Don't be confused by the word "modular" here, as we are referring to the fact that some of these devices snap together with different options.

Fixed jack plates are the traditional type of plate. These plates are made of molded plastic and include the opening for the modular connector as well as a mounting for the connector and termination components. The most common fixed jack plates have one or two connector positions and are often referred to as *simplex* or *duplex* jacks (respectively). The plate opening and its related molded structures form an integral part of the connector assembly. The connector cannot be separated from the plate and used alone. This is the reason the jack plate is called fixed. The supports for the connector terminations are also usually molded into the plate. The most common type of termination is a screw and washers that attach to a molded post on the back of the plate. An example of a fixed jack plate is shown in Fig. 7.5.

With the advent of structured wiring and the tough requirements of Category 5, modular jack plates and their associated connectors have become much more common than fixed jack plates. These modular plates have simple (usually square) molded openings into which are snapped a variety of connector modules. The advantage to this modular approach is that the connector style, terminations, and even color are independent of the mounting plate. Figure 7.6 shows the modular jack plate and connectors.

Plates are available to hold from one to six connector modules, so you can mix and match connector options. This means that you can decide to mix 6 and 8 position jacks on the same plate and in any order. Some of the modular systems even have modules for coax and fiber. You could have a phone jack, LAN jack, CATV jack, and fiber optic jack all on the same plate. Alternatively, you could have phone and LAN connections for two workstation positions all on the same jack plate. While the standards make no requirement for multiple module outlets, they certainly do not prohibit it. Just be careful to include a larger outlet box behind a plate with several connectors. You will need a space to coil up all that cable.

Figure 7.5 An example of a fixed jack plate.

The connector module includes the modular connector jack and the termination mechanism for connecting the cable. Termination is accomplished by several types of IDC connectors, as we will describe later. The construction of the connector/termination module generally produces much lower noise and crosstalk than with conventional fixed jack plates. For this reason, these jacks are preferred for LAN wiring, even when making a Category 3 installation. They are also much easier and quicker to terminate than conventional screw terminals. Notice we did not say that they were less expensive.

Another variety of jack plate is a two-piece modular jack assembly. These assemblies are unique in that they separate the wire termination function completely from the connector function. A typical two-piece assembly will include a backplate module that the wire terminator snaps into after the cable is terminated. The terminator module is essentially a special type of connector that mates with a printed circuit board-edge contact area. This arrangement is commonly called an *edge connector*. The back module then mounts into the outlet box using the normal mounting screws. To finish the assembly, one or more connector modules, each with a PC-board edge contact, plug into the terminator module at the rear of the back module. A faceplate is then attached to secure the connector modules in place. This type of assembly has the unique feature that the front connectors may be changed without reterminating the cable. For example, you could change from a single 8-position connector to a dual 8-position one, or change to a 6-position or even coax module. Of course fiber connectors are out, because it is cable that is terminated, not fiber. Figure 7.7 shows an example of a two-piece modular jack assembly.

Modular jacks are often referred to as "RJ-45" jacks. This is not really a correct moniker, although it is in very common use. Because all modern LAN wiring uses the 8-pin (8 position, 8 conductor) modular plug and jack, we will call these connectors "8-pin modular" or just "modular." For a more complete discussion, see the selection "RJ-45, What's in a Name" in Chapter 10.

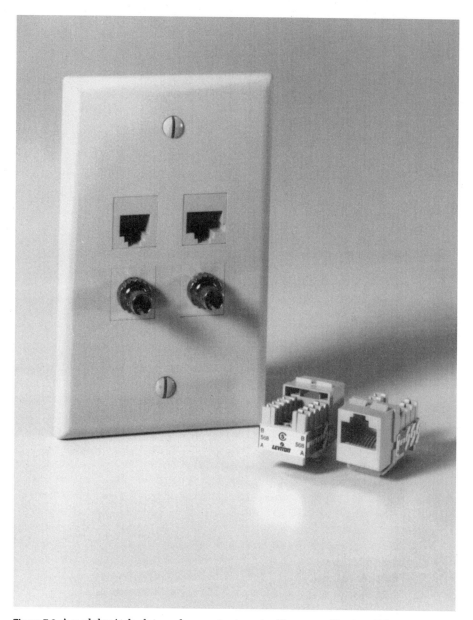

Figure 7.6 A modular jack plate and connector inserts. *(Courtesy of Leviton Telcom.)*

Figure 7.7 An example of a two-piece modular jack assembly. *(Courtesy of Amp Incorporated.)*

Termination types

Two basic types of wire termination are used with modular jacks for twisted pair wire. These are the screw terminal and the insulation-displacement connector (IDC). The screw termination is the original type that was used with telephone systems. It is still found in some LAN wiring installations, but is not recommended for Category 4 or 5 operation. As a matter of fact, you probably will never see screw-terminal connectors certified above Category 3, if they are certified at all. You definitely do get a better connection, from the viewpoint of LAN transmission, with the other type of termination. For that reason, many installers do not recommend screw terminals for any LAN work. Of course, you do not need to rush around and pull out all of your old screw terminal plates that might be in your network now. Just remember that they cannot perform above Category 3 and that old installations may have "space wired" station cable that is untwisted or crossed before it even gets to the screw terminals.

Nevertheless, there is still a right and wrong way to terminate wire on screw terminals. Proper wire termination for screw terminals is illustrated in Fig. 7.8. The screw terminal consists of a screw, two washers, and the plastic post of the jack plate. The jack's connector wires are normally terminated in an open-end spade lug that is positioned between the plastic post the lower of the two washers. To terminate the station cable, the screw is first loosened. Each station wire's insulation should be stripped back about ⅜-inch and the bare copper portion wrapped between the two washers. The wire may be preformed before it is placed into position. The screw is tightened moderately and electrical contact is made in the sandwich made by the washers and the plastic post. The proper direction for the wrap is clockwise around the screw (and it must be between the two washers). No insulation should be between the washers and the wire should not wrap back across itself (to avoid breaking the wire when the screw is tightened). Some installers strip an inch or more and wrap the wire around the screw (with no overwrap). The excess wire should be trimmed after the screw is tightened. It is also a good idea for LAN circuits to try to maintain as much of the pair twist as possible, up to the point of termination. Some authorities recommend trying to limit the untwist of Category 3 cable to no more than 1.5 to 2 inches.

The other type of wire termination is the IDC contact. This contact avoids stripping the insulation manually, by cutting through the insulation during the termination process. The wire, cut to length, is inserted into the contact area and snapped into place, either by a special tool or by snapping a plastic cap onto the IDC contact. Either action causes a small area of insulation to be pierced and electrical connection made with the underlying contact. There are three common types of IDC contact mechanisms: the 66-type, the 110-type, and the snap-in type. Other proprietary mechanisms include BIX and Krone types.

An older type of IDC jack contact field is fashioned from a shortened version of the traditional 66-type punchdown block. In this jack type, the wires of the modular connector are connected to positions on the block into which wires from the station cable are punched. The amount of untwisted wire from the block to the connector is usually too great for this type of jack to be certified above Category 4, if indeed it is certified at all.

Insert wire between washers
(clockwise wrap)

Insert spade lug below washers
(above post)

Figure 7.8 Proper wire termination for screw terminals.

Another type of IDC jack contact incorporates the newer 110-type punchdown strip as shown in Fig. 7.9. The 110 strip may be separate from the modular connector and connected by wires, as with the 66-type jacks, or it may be an integral part of the connector module. If discrete wires connect the strip, the same untwisted-wire problem exists as with the 66 blocks, and certification above Category 3 or 4 is rare. However, if this block is incorporated into the connector module, the combination can be certified to Category 5. The 110 strip has the advantage that it can be terminated with conventional punchdown tools and its connectivity and longevity parameters are well established. On the other hand, the strip is often wider than the connector module it is attached to, which limits placement on a mounting place. Variations of this scheme use the BIX or Krone blocks, both described in Chapter 8, in place of the 110 block.

An alternative to these two other IDC connectors is what we are calling the "snap-in" connector for lack of a better name (see Fig. 7.6). This connector type is made in a number of styles by many manufacturers. In general, these designs all have the common characteristic of compact construction and toolless termination. The connector contacts mount directly onto a printed circuit board,

Figure 7.9 A connector jack that used 110-type blocks for the horizontal cable connections. *(Courtesy of Molex Fiber Optics, Inc.)*

which also mounts the modular connector. The wires are each cut to size and placed in a channel above the contact edges. A plastic insert is used to push up to four individual wires into the jaws of the IDC contacts, where the insulation is pierced and the electrical connection is made between the wires and the contacts. The insert usually snaps into place with a click, which is why we are calling it a *snap-in connector*. The beauty of this connector style is that it terminates the wires without any tools. Its compact design leaves room on the connector housing for color codes to match the wire colors. Pin numbers may also be printed on the housing to help the installer.

Standard jack pin-outs

The standard pin connections for a variety of modular jack styles are shown in Fig. 7.10. Modular jacks are used for many different purposes, of which LAN wiring is only one.

The TIA 568-A standard describes two different wiring patterns for wiring the 8-pin modular jack. The two patterns are called T568A and T568B, and are shown in Fig. 7.10. The two patterns differ only in the placement of Pairs 2 and 3 of the traditional 4 pair order. This order is defined as Pair 1 = Blue, Pair 2 = Orange, Pair 3 = Green, and Pair 4 = Brown. The T568A pattern places Pair 2 on Pins 3 and 6 of the jack, and Pair 3 on Pins 1 and 2. The T568B pattern places Pair 2 on Pins 1 and 2 of the jack, and Pair 3 on Pins 3 and 6. The A-pattern is actually unique to the TIA standard, but it is now the recommended pattern. The B-pattern is the traditional AT&T 258A pattern specified in their premises distribution system. Ironically, the A-pattern better follows the progression of wiring patterns from the 2 pair pattern for 6-pin modular jacks that is used with many telephone sets (the RJ-11/12 style jack). By the way, the "-" is used in TIA 568-A specifically so there is less confusion with the wiring pattern called T568A.

There has been considerable controversy as to which pattern is the proper one to use. Depending upon whom you ask and where you ask, you can get strong opinions on either side of the question. The TIA standard, however, is quite clear. The "pin/pair assignments shall be..." T568A. Optionally, the T568B may be used, but only "if necessary to accommodate certain 8-pin cabling systems." These certain systems are ones that already use the optional pattern, for the most part. If you do not need to connect to an existing system, cable, patch, etc., that uses T568B, you should use the recommended T568A pattern.

What difference does it make which pattern you use—T568A or T568B? Electrically, it makes no difference, as long as exactly the same pattern is used at *both* ends. Let's say that again: the two patterns are electrically equivalent, as long as both ends of a cable run are wired with the same pattern. As has been said, "the electrons don't know the wire color." This means that as long as Pin 1 and Pin 2 on one end of a link connect to Pin 1 and Pin 2 on the other end, and the same is true of Pins 3 and 6, you can use either pattern. Both

T568A

Ethernet 10/100BaseT

T568B / AT&T 258A

Token Ring

USOC RJ-45S

USOC RJ-48S / TP-PMD / DDS
(For T1, see note 4)

USOC -8 (RJ-61)

Arcnet

Pair Name	Conductor	Color Code (horiz. wire)	Solid Color (internal jack)
Pair 1	T1	White/Blue	Green
"	R1	Blue/White	Red
Pair 2	T2	White/Orange	Black
"	R2	Orange/White	Yellow
Pair 3	T3	White/Green	White
"	R3	Green/White	Blue
Pair 4	T4	White/Brown	Orange
"	R4	Brown/White	Brown

Notes:

1. T = Tip, R = Ring.
2. When polarities are used,
   Tip is +, Ring is −.
3. Jacks are viewed from the front.
4. T1 uses the RJ-48C and RJ-48X
   variations: Pair 1 is pins 4 & 5,
   pair 2 is pins 1 & 2).

Figure 7.10 The modular jack may be used for a variety of purposes.

the outlet and the wiring closet terminations (punchdown block or patch) of a horizontal cable run should be wired with the same pattern, as should both ends of any patch or user cord.

Technically, you can use either pattern; however, some patches, fanout cables, equipment connectors, and adapters are prewired with the T568B pattern and you must match that pattern at the other end of the cable for the link to work. You can, however, mix patterns from one cable run to the next within the same link. For example, a patch cord could be wired with either pattern without regard to how the horizontal cable was terminated.

A common wiring pattern problem can occur in the horizontal crossconnect. If you terminate the horizontal cable into a punch block first, then crossconnect to a patch, you should be sure that the patch wiring matches that of the corresponding wall outlet, and that you have preserved the pair order during punchdown and crossconnect. See Chapter 9 on patch panels for more details on the internal wiring pattern of a patch panel.

The official stance of the TIA is that the A pattern is preferred. That statement certainly applies to new installations, since it would be problematic to switch the pattern in an older installation. You could also justifiably choose the B pattern if AT&T components are used, since that is their standard offering. There is great debate over which pattern is used the most. The truth is in the hearing. The customary pattern varies widely, depending upon whom you ask and what part of the country (or world) you live in. However, it seems that the A pattern originally was not used as much because it was different from the pattern that had historically been required for telephone use. Now that most manufacturers offer both, you are free to follow the TIA's preference, or yours.

What happens if you use one pattern on one end of a cable run and another on the other end? Most LAN network connections will fail, and telephone connections that require two pairs will also fail. Pairs 2 or 3 are used by almost every device we connect to an 8-pin modular jack.

The TIA wiring patterns are considered universal in that they are applicable to several types of connections, because they cover all eight pins of the modular connector. For example, the TIA patterns can be used for any of the Ethernet connections, such as 10 and 100BaseT. The TX and T4 variations are both handled, as is 100VG-AnyLAN, since all four pairs are specified. In addition, the TIA patterns can support Token-Ring, Arcnet, ISDN, T1, ATM-PMD variations, and, yes of course, telephone wiring in its many variations.

You can make a very good argument that the TIA 568-A wiring system is as close to a universal wiring system as we will ever get. Of course, there will always be situations that do not fit this standard model, but more and more equipment is being designed to take advantage of its characteristics. If you choose to diverge from the standard, to implement some of the 2 pair technologies, for example, you may be left out in the cold in a very short time. For example, the only practical way to increase data rates on unshielded twisted pair wiring beyond the current 100 Mbps data rates will be to use additional pairs in a multiplexing or encoding scheme. The 100BaseT4 and 100VG-AnyLAN standards do this now with encoding such that the actual pair bandwidth requirements are under 32.5 and 15 MHz, respectively. This means that a future standard could go well beyond the current top data rates without having to replace installed Category 5 cable links, which are rated to 100 MHz. However, these new ultra-high speed technologies will very likely need all four pairs. You cannot very well use four pairs if you decided to wire just two, just to save a little money in the short term. It is therefore strongly recommended that any new installations adhere very closely to the TIA standards, including the number of pairs to each outlet jack.

TIA Performance category

It is very important to fully understand the categories of operation of the various modular jacks. The TIA 568-A standard contains strict performance requirements for the outlet connectors that are used for Category 3, 4, or 5 operation. The requirements differ somewhat by category, although Category 4 and 5 parameters are often the same. Although Categories 1 and 2 exist, they are not the subject of the standards because they really are not appropriate for any type of LAN cabling.

Components, as well as cable, should be marked as certified for a particular category of use. Thus, a connector that is intended for Category 5 use should be marked for Category 5, as explained below. However, you may have occasion to need to determine the appropriate use of an unmarked jack. While it may not be possible to categorize an unmarked jack without some uncertainty, there are some guidelines that can help out. If you can identify the jack assembly style, you should be able to determine its probable category of rating. The jack styles and their appropriate categories for use are shown here:

TIA performance category	Termination style
1-3	Screw terminal or IDC
4	IDC, including 66- and 110-type
5	IDC using PC board, only*

*Category 5 jacks can also use 110, BIX, and Krone type connections.

Of course, any jack that is rated for a higher category than the intended use can be substituted. For example, it is perfectly acceptable to use a Category 4 or 5 jack for a Category 3 installation. However, you could possibly upgrade the entire wiring system for very little more cost. The IDC jacks are so easy to use, they are often used even when their higher category of operation is not required.

Keyed-plug entry

Some modular jacks have an extra slot at the side to supposedly limit connections to cable plugs with a matching slot. These side keyed jacks are largely useless because they easily accommodate a nonkeyed plug. The original intent was to protect delicate digital equipment from the higher voltages that can sometimes be present on telephone circuits. For example, the common open circuit voltage for a telephone line in the United States is -48 V dc, and the ring voltage can approach 90 V ac (both at low current). However modern data circuits are fairly immune to these voltages and the fact that the keyed jacks are easily defeated has led to their disuse.

An alternate style of keyed jack is the MMJ style used by Digital Equipment Corp., described in Chapter 5. This jack has primarily been used for the connection of RS-232 and RS-422 low speed devices and is rarely found in other uses. It is not appropriate for standards-compliant LAN wiring.

Fiber Optic Outlet Jacks

Outlet jacks for fiber optic connections have several unique requirements. The fiber optic outlet must allow the station cable fibers to be terminated and connected into a jack plate. In addition, the design of the jack plate must allow the easy attachment of an optical jumper to the workstation equipment. A special passive coupling device, called an adapter, is used to connect the two fiber optic cables at the outlet jack.

Unlike the termination of metallic conductors, the termination of an optical fiber can be a delicate and time consuming process, although the time may be reduced by using more expensive "quick-termination" connectors. The fiber must be stripped of its protective coverings, including the binder. It must then be precisely cleaved before being inserted into the mechanical coupling that is loosely called the optical connector. Some fiber connectors require the use of an epoxy or glue to secure the fiber and may use a crimp or clamp-in arrangement in addition (or in lieu of) the glue. Finally, the connectorized fiber end must be carefully polished to a smooth surface so that light entering or exiting the fiber is not bent off-axis (the polishing step is not required for a quick-termination connector).

In order for light to couple properly between connectors, the internal fibers must be carefully aligned and placed virtually in contact by the coupling device. In fiber optics, this coupling device is called an *adapter* rather then a connector, because it does not actually terminate any fibers. It mounts in the fiber optic outlet plate much as a coax bulkhead barrel adapter does in coax systems. The fiber adapter provides a keyed location where the terminated station fiber connects on the back side of the outlet plate. The plate is then mounted in the fiber optic outlet box. The fiber optic user cord is connected to the section of the adapter that is on the front side of the outlet plate. Because of the nature of fiber optic transmission, two fibers and connectors are generally required for each LAN connection, one for transmit and one for receive. Keying these two connections, or differentiating them by color, is used to orient the transmit and receive fibers.

Fiber optic connectors

There are several types of fiber optic connectors. We will mention three of these that are often used in LAN wiring. The first is the optical SMA (SMA-905/906) connector shown in Fig. 7.11. This connector is quite similar to the miniature, screw-on SMA connector used for radio-frequency connections on small-diameter coax cables. The fiber is contained in a metal or ceramic ferrule and actually comes through to the very tip of the ferrule. The SMA connector allows very precise alignment of the fiber, but has the disadvantage of being awkward to use. To connect the SMA, the ferrule must first be aligned with the mating connector, and then the outer threaded retaining sleeve must be screwed down to hold the fiber in place. The 905-type connector has a straight 3 mm ferrule. The 906-type has a step-down ferrule that allows two ferrules to be connected by a coupler, with a plastic sleeve to align them.

SMA connectors are not designed to be easily paired together to support the standard dual fiber interface. You must mate each fiber connector one at a time. Also, there is no easy way to key this connector so that the transmit and receive fibers do not get interchanged. Although this connector was once the most common fiber optic connector for routine LAN connections, it is not in as wide use today.

Figure 7.11 The optical SMA (SMA-905/906) connector. *(Courtesy of Molex Fiber Optics, Inc.)*

Another popular fiber optic connector is the ST connector, shown in Fig. 7.12. The design of this connector has a bayonet-style outer sleeve that allows the connector to be quickly inserted in its mate and secured with a quick quarter-turn. The connector looks somewhat like a miniature BNC connector. The ST connector has begun to supplant the SMA connector in most equipment and chassis connections. However, it still has the same single-connector problems that the SMA did. The connections must be made one at a time and there is no practical way to bind two fiber connections into one unit.

A newer connector is called the SC. It allows quick connections, good fiber alignment, and multifiber coupling so that two may be combined into a duplex connector for transmit and receive. The SC connector is an unusual looking square tipped fiber optic connector that is slightly cone-shaped at the tip, as shown in Fig. 7.13. The SC is a push-on, pull-off connector that uses spring retention to hold the connector in place when mated. The square design and push/pull coupling mechanism make the connector usable in high density applications. The SC connector may be coupled in duplex sets with a coupling receptacle or duplex clip. In a keyed duplex set, the SC connector easily implements a form of polarity matching with coupling adapters in fiber optic outlets or patch panels. This connector has been adopted by the TIA as the officially recommended fiber optic connector in TIA 568-A. The connector is called the 568SC connector in the standard.

Which Fiber Connector?

The advent of fiber optic cable communications systems has seen the progressive introduction of a wide variety of optical connectors. The dominant types that have been used for LAN connections are the SMA, the ST, and recently the SC. Which connector should you use?

All of the three connectors mentioned provide close fiber alignment, reliability, and some measure of industry standardization. However, the SMA and the ST are not easily adaptable to a duplex arrangement which allows two fiber connections to be made at the same time. Both these connectors styles must be used individually, because they require the twisting of a threaded or a bayonet locking ring to secure the connectors to the mating jack. The SC connector, on the other hand, uses a push-on/pull-off mechanism which allows the two individual connectors to be attached together and connected or disconnected simultaneously.

The SC connector has been selected as the required connector for TIA 568-A compatible installations. As a matter of fact, the connector is designated "568SC" when mounted as a duplex connection with 12.7 mm (0.5 in) ferrule spacing. The standard suggests an upgrade strategy for existing installations and, to be compatible with the standard, you must adapt older connectors and use the 568SC on new installations.

Unfortunately, the SC connector is quite new and, until recently, proprietary. Thus very few manufacturers currently offer equipment connections with the SC connector. The answer to this compatibility problem is to adapt existing equipment with SMA, ST, or other types of connectors to the SC system. This is done by either using an adapter cable, with the 568SC on one end and an SMA or ST on the other end, or by using a hybrid adapter coupling, which couples an SC on one side to an SMA or ST on the other side. Either solution should cost about the same as the standard cable or adapter.

Which fiber connector should you use for a new installation? SC, of course.

Figure 7.12 Another popular fiber optic connector is the ST connector. *(Courtesy of Molex Fiber Optics, Inc.)*

Figure 7.13 The SC connector is an unusual-looking, square-tipped fiber optic connector slightly cone-shaped at the very tip. *(Courtesy of Molex Fiber Optics, Inc.)*

Other Outlet Jack Types

IBM cabling system jacks

A few other outlet jacks deserve mention. The first is the IBM Data Connector. The Data Connector, shown in Fig. 7.14, is a hermaphroditic (genderless) connector designed so that any connector mates with any other. It is a rather large, square-nosed connector with a rectangular body. The Data Connector is designed to be used with shielded twisted pair (STP) cable, although it can also be used with other connectors.

The Data Connector is now specified in TIA 568-A as the standard connector for use with STP-A cable. The use of STP with the Data Connector was previously specified by an *EIA Interim Standard Omnibus Specification, NQ-EIA/IS-43.* The TIA 568-A standard now details the Data Connector and its use with STP-A shielded cable. The Data Connector's complete electrical and mechanical specifications are contained in *IEC-807-8, Rectangular Connectors for Frequencies Below 3 MHz, Part 8: Detail Specification for Connectors, Four-Signal Contacts and Earthing Contacts for Cable Screen, First Edition.* Although this connector specification states that the connectors are for 3 MHz and below, the TIA has extended their characteristics to 300 MHz.

The color codes for horizontal 150 ohm STP-A cable are:

Pair 1	Red, Green
Pair 2	Orange, Black

This color code corresponds to the markings for the IDC contacts within the Data Connector and makes the cable conductors very easy to match with their proper contact.

The Data Connector outlet is a version of the standard Data Connector that is mounted in a cover plate. The cover plate is attached to a standard outlet box. The STP-A station cable is terminated directly in the outlet connector. There are versions of Data Connector faceplates available that contain both the Data Connector and one or more standard modular jacks.

The Data Connector and STP-A cabling system is a robust system that offers certification to 300 MHz, three times the current maximum frequency specification of standard UTP systems.

Coax jacks

Coax jacks are another type of outlet jack. The BNC coax jack is commonly encountered in 10Base2 Ethernet cabling systems. This cabling system calls for the thinnet segment cable to be run from one workstation position to the next, and so on until you reach the end of the segment. At each workstation location, the two cable sections are joined together by a BNC-T and the center position of the T is connected to the workstation network adapter. The workstation essentially taps the coax line to become a part of the network. The

Figure 7.14 The Data Connector is a hermaphroditic (genderless) connector designed so that any connector mates with any other. *(Courtesy of Thomas & Betts Corporation.)*

method generally does not allow an extra stub of coax from the tap to the workstation; both cable sections must be extended all the way to the T at the workstation adapter.

Each fixed 10Base2 coax outlet, thus, needs two coax connectors, one for each cable section leading to the workstation. The customary outlet has two separate female BNC connectors on the faceplate, each connected to one of the two cable sections at the rear of the faceplate. There are two methods to connect the cable sections to the rear of the faceplate. One method connects bulkhead-mounted female connectors directly to each cable end. Each connector is inserted through a d-keyed hole in the plate and secured with a thin washer and nut. The alternative method places two bulkhead-mount barrel connector/adapters in the holes of the plate. The coax cable ends in the wall are terminated with male BNC connectors, which are then connected to the back side of the bulkhead adapters. From the front, the plates look the same, regardless of the method you use to connect the cable ends.

From the dual coax outlet, connection to the workstation is made with two coax jumper cables, with two BNC male connectors on each cable end. At the workstation, a BNC-T is connected to the network adapter and the two cables connected to the arms of the T. If the workstation is at the end of the coax segment, only one cable is used and a 50 ohm terminator is placed on the other T position. If there is no workstation in place, a short coax jumper cable is used to connect the two bulkhead connectors on the outlet's faceplate, in order to complete the cable circuit to the next workstation. If the workstation position would have been at the end of a coax segment, a single terminator may be used instead. The 10Base2 system requires end-to-end connectivity and dual-ended termination of the coax segment for proper operation. Any disconnected cables or terminators may cause the entire cable segment to cease to function.

Marking Outlets

Outlet jacks should be clearly identified on the face plate. The identification should include a cable number and jack number, as well as an indication of whether the jack supports data (such as a LAN connection) or telephone. Some jack inserts and faceplates are marked to indicate "data" or "phone." Telephone extension numbers generally should not be marked on the plate, since many phone systems allow the number to be assigned to any physical outlet. The same applies for data jacks that correspond to a particular port number on a LAN hub or other network connection. You should keep a chart to match each extension or port number to each jack number.

The scheme in EIA/TIA-606 is a good basis for marking the jacks. The markings should be clearly visible with bold lettering about 3/8 inches in height. The standard calls for marking at the front of the jack position, above or below the jack. However, you might also want to consider a neat marking on the top of the plate, so it can be read from above. (If you ever have to peer down behind a desk or a cabinet to identify an outlet, you will be thankful for the top markings.)

The outlet markings should identify the cable number, and, in a large facility, the wiring closet as well. For example, a marking for wiring closet number 3, cable 42, might be W3-42, or W3-J42 to differentiate it from a cable number. It is best if the outlet number corresponds to the cable number in the wiring closet. The cable itself can also be marked with a variety of vinyl labels, but the type of label that is part of a tie-wrap should be avoided on Category 5 cables, for the reasons previously mentioned. If all else fails, fine-point indelible markers will also effectively mark cable numbers on a cable jacket. Station cables are usually either in the wall or at the rear of termination blocks in the wiring closet, so the marking doesn't have to be pretty. Neatness counts, however, as you must be able to read the cable number later.

Outlet jacks are often color-coordinated with the room decor, which is fine as long as standard NEMA electrical colors are chosen. Nonstandard colors may cause a clash with other electrical outlet plates and may be difficult to obtain. If the electrical plates have already been selected, you should try to match those colors.

Modular outlet jacks offer a variety of colors for the modular inserts. In theory, the color markings allow one to easily differentiate between, say, voice and data jacks. However, there are really no universal standards for these colors. What may seem to be a logical color choice to you could be the opposite of what someone else might choose. You might simply want to leave all the colors the same neutral color as the faceplate and differentiate the jack functions with the plate markings. Alternatively, you may choose to leave the telephone connector neutral and use a differing color for the LAN jack.

Workmanship

The subject of workmanship is particularly important with regard to jack termination. The higher categories of wiring have strict guidelines for the amount of untwist that may be present at any point of termination. When a pair of wires is untwisted, even slightly, it can cause a greater disruption to

Figure 7.15 A sample of a cable link with one improperly untwisted connector termination. The resulting NEXT anomaly is shown in the scan graphic.

the high-bandwidth signal transmission than almost any other noncontinuity defect. Furthermore, there are several points of unavoidable untwist in any cable link, and the effects of untwist are somewhat additive. You should carefully inspect at least a sampling of cable terminations to make sure they are done properly.

Figure 7.15 shows a sample of a cable link with one improperly untwisted connector termination. You can see that there is a very large cable anomaly at approximately 72 feet from the tester. This test link was made up with an intentional excess untwist of ¾ inch at one of the connectors of the 1 foot patch cord. When you compare the size of this anomaly to the other connectors (which are all properly made), you can clearly see the reason we emphasize good workmanship. Patch cords are easy to inspect visually, but your outlet jack wiring is hidden in the wall. Be sure that the person who installs your wiring is properly trained to do these connections well and test each horizontal cable run with a cable scanner.

Another workmanship issue that affects station outlets is the amount of excess cable that is left at the outlet. Standard practice is to leave no more than 18 inches of extra cable length when wiring the outlet. This distance is measured from the wall opening to the jack before it is installed. Excess cable should be trimmed before the outlet is terminated. You should leave a reasonable length of extra cable to simplify the initial termination and any subsequent inspections or reterminations of a cable run.

8

Wiring
Closet Termination

Another important area in the chain of wiring devices that makes up the LAN wiring connection is the cable termination in the wiring closet. The wiring closet is the point of termination of all of the cables that go to the workstation areas and the cables that interconnect to other wiring closets. The wiring closet will contain the punchdown blocks, patches, and other termination and wire-routing devices necessary for the interconnection of horizontal and backbone cabling. It may also contain various types of equipment, such as wiring hubs, routers, and even network servers. Other names for the wiring closet covering its more specific uses include telecommunications closet, main cross-connect, horizontal cross-connect, intermediate cross-connect, equipment room, and entrance facility.

In the generalized model of universal, structured cabling, the wiring closet also may contain telephone switching and distribution equipment. Much of what we cover in this chapter will be equally applicable to telephone wiring, but we will emphasize the LAN wiring aspects wherever possible. It is interesting that the wiring closet is one place where the differing needs of telephone and LAN wiring are most apparent.

The very strict length and workmanship guidelines that are required for a successful LAN wiring system are very much relaxed if you are installing telephone wiring. For example, telephone wiring is often run through multiple sets of punchdown blocks in several wiring closets with little regard to total distance or care in routing. (Although the TIA standards do have strict guidelines, they are often excessively strict for telephone connections.) The much lower audio frequencies of most telephone installations allow these signals to run thousands of meters on very loosely-twisted pairs. However, if you intend for your wiring system to be general purpose, so that you can run either voice or network signals, you must design your system to meet the more stringent operating rules embodied in the TIA standards. LAN wiring requires strict adherence to length, routing, rating, and workmanship rules in order to operate properly. The extra requirements for LAN operation are really the determining factors in the design of a modern structured wiring system.

In this chapter, we will cover all of the types of termination and routing devices, other than patch panels, that are used in the wiring closet. Patch panels, jumpers, and cross connects are covered in the next chapter. This chapter will also cover some of the mounting and location issues of the wiring closet, including the options for locating hub equipment.

Punchdown Blocks

The fundamental component for wiring termination in the wiring closet is the punchdown block. The punchdown block can take many forms, and has over the years evolved into a rather complex "system" component. The punchdown can even be incorporated into a patch panel, but we will talk about that later.

There are two main types of punchdown blocks in common use: the 66M block and the 110 block, both originated by AT&T. Both types are now offered by a number of companies and versions of these block terminations are incorporated into many products, including outlet connectors and patch panels. These two punchdown types dominate the installed and new markets, so we will cover them in some detail. We will also cover a couple of proprietary punchdown systems that are available from only one manufacturer or are not as widely used.

Type 66M connecting blocks

The original workhorse of telecommunications termination is the type 66 connecting block, shown in Fig. 8.1. This style of termination block has been around for decades and its use in the telephone industry is pervasive.

There are several types of 66 blocks, but the most common is the 66M1-50. This block has 50 horizontal rows of wire termination contacts, with four bifurcated contact prongs in each row. (*Bifurcated* means the contact is split in two, so the wire can be held in place by the bifurcated fingers of the contact.) Each contact unit is called a *clip*, and the contact clips in this style of block each have two prongs, stamped from the same piece of metal. The four contact prongs in each row are paired 1-2, 3-4, with each pair of contacts electrically and mechanically connected. Figure 8.2 shows a cutaway view of one contact row. Some varieties of the 66M block have four common contacts in each row, while others have four totally independent contacts in each row, so be careful what you use.

Connectorized 66 blocks with 50-pin connectors are also available, as shown in Fig. 8.3. These blocks are usually manufactured with clips that have a wire-wrap post protruding from the bottom of the clip. When the assembly is manufactured, a wire is wire-wrapped to the post on each clip with the other end connected to a 50-pin "telco" connector mounted on the side of the bracket. The connector may be either male or female. The contact pins of the connector are

Figure 8.1   The original workhorse of telecommunications wiring termination is the type 66 connecting block. (*Courtesy of The Siemon Company.*)

typically wired to all 50 rows of contacts, with one pin on the telco connector corresponding to each row of the block. Connectorized 66 blocks are available with one or two telco connectors. Single-connector blocks are connected to only one column of paired contacts, while dual-connector blocks are connected to

Figure 8.2 The four contacts of a 66-block contact row are paired 1-2, 3-4, with each pair of contacts electrically and mechanically connected.

both, one on each side of the block. The connectors may be used to attach pre-assembled 25 pair jumper cables or to attach a modular fanout (octopus) cable.

The fanout cables, if used, come in several varieties, with 2, 3, or 4 pair modular legs (so they are often not "octo" cables at all). The 4 pair variety fans out to cable six legs, each terminated in an 8-pin modular plug. The legs may be ordered to length, but the wire should be stranded twisted pair wire and rated to the appropriate category in order to conform to LAN wiring guidelines. Because the 66M has 50 contact rows, and a 4 pair leg has eight wires to terminate, and eight divides into 50 rows only six times (evenly), there are thus six legs from the fanout when using 4 pair cable legs. The last two rows of the 66M are not needed in most pairings, so only 48 of the 50 rows are used. Similarly, the 3 pair fanout has eight legs ($48 \div (2 \times 3) = 8$), and the 2 pair fanout has 12 legs ($48 \div (2 \times 2) = 12$). The modular plugs of fan-out cables may be wired with any of several wiring patterns, including T568A and T568B. When ordering these fanout cables, you should specify which pattern you want, and ensure that the corresponding outlet jack (at the other end of the station cable) is wired the same. See Chapters 7 and 10 for more information.

Horizontal station cables (or backbone cables) are routed down the mounting backboard underneath the 66 block for termination. They are threaded out through the openings in the side of the mounting bracket where they are to be terminated. The outer jacket of the cable is removed to expose the wire pairs. The pairs are fanned out, sorted by color, and routed to the appropriate contact. Each wire passes through one of the narrow slots in the face of the block. The wires generally enter the slot that is just above the target contact row and are wrapped down into the jaws of the contact for termination. The contact has a hook on one prong to help position the wire until it is punched down. The side slots hold the wires in place-order, although they do not in any way support or strain-relieve the wires. (By the way, the process of fanning out, sorting, wrapping, and terminating wires on a block is called *stitching*. And one *punches down* the wire with the impact tool, equipped with an appropriate blade for 66 block termination.)

The 66 block has a TOP marking to indicate the way it should be positioned. If it is connectorized, the connector pins will be numbered upside down if the block is accidentally reversed. If the block is not connectorized, mounting it upside down just looks amateurish (and you will get comments). You should always punch down all the wires of a station cable, even if you do not intend to use all of them immediately. It is impractical to later use the extra pairs of a cable if they are not already punched down, because there is nowhere to terminate them. If for some reason you must install a cable that has more pairs than you generally use (such as a 4 pair cable in an installation that otherwise uses 3 pair cables), then simply cut the extra pair(s) flush at the point where the cable jacket was removed. You may be able to pull the jacket back slightly before you cut off the excess pair(s). Then push the jacket back in place to make a neat installation.

The 66M block is designed to terminate unskinned *solid* copper conductors with plastic insulation. Conductors from AWG 20 to 26 can be accommodated. The use of stranded wire is not recommended under any circumstances. To terminate a wire, the wire is pressed down into the slot between the fingers of the contact with a punchdown impact tool, shown in Fig. 8.4. The tool causes the insulation to be displaced, leaving the copper wire and the phosphorus-bronze clip in direct,

Figure 8.3 Connectorized 66 blocks with 50-pin connectors are also available. *(Courtesy of The Siemon Company.)*

Figure 8.4 To terminate a wire, the wire is pressed down into the slot between the fingers of the 66 block contact with a "punchdown" tool. The optional 110 blade is also shown. (*Courtesy of The Siemon Company.*)

gas-tight contact. The spring action of the tin-plated contact holds the wire in place. The tool also trims off the excess wire after termination. *Only one* conductor may be terminated in each contact slot. If a wire is removed from a contact slot, the slot may be reused, but only if the new wire is the same or larger size. Using a smaller wire, using a slot repeatedly, or punching two wires into a slot will cause the connection to fail, sooner or later.

As we have said, station wires are punched down from both sides of the block. Because the block has 48 usable row positions, that means that you can terminate 6 station cables per side of the block, or a total of 12 cables per block, assuming 4 pair cables. On a standard (nonconnectorized) 66M block, you should punch down the station wires on the outer contact of each dual pronged clip. (Some connectorized blocks wire to the outer contacts, rather than using a wire-wrap clip.) Crossconnect wire pairs are introduced to the row in the same way, through the slot just above each contact row, but punched down on the inner contact of each clip. The crossconnect wire overlaps the station wire that was punched into the outer contact, but the color code of each should still be visible. When punching, take care that the punchdown tool is turned the correct way so that the excess wire, and not the intended connection, gets trimmed off.

The tendency to fan the wires out, and the amount of untwist that exists where the wires of a pair pass through the side slots, creates a problem for higher wiring categories, such as Category 5. The contact clip also is a rather large piece of metal that can contribute to impedance mismatch and crosstalk at higher frequencies. For that reason, standard 66M blocks are not suitable for use above Category 3.

Some manufacturers have introduced low crosstalk 66 blocks that are rated to Category 5. The standard wiring technique of one-wire-per-slot can still cause an unacceptably high amount of crosstalk, even with these special blocks. If special 66 blocks are used for Category 5, the wire pair should be inserted intact through the side slot between the two target contact rows, rather than split and run through two side slots. The wires are then wrapped up or down (whichever is appropriate) into the contact and terminated. This method maintains the amount of untwist well under the ½ inch maximum. Unfortunately, this technique also makes reading of the color stripes on the wires difficult after termination, so you should be very methodical when wrapping the wires to be sure you do it right the first time. A subsequent termination on the inner contact will make the color stripes of the wires on the outer contact of that row almost impossible to discern.

Many accessories, including adapters, bridging clips, and plug-on jacks, exist for the 66M block. Virtually all of these are inappropriate for use in permanently installed LAN wiring systems. The most stable connection system for LAN wiring is to use crossconnect wires between punchdown blocks and patches, or to terminate directly on the patch and skip the punchdown block altogether.

A version of the 66 block incorporates 8-pin modular jacks into the assembly. The jacks are mounted in groups of four or so at the sides of the mounting bracket. This type of block may eliminate the need for a separate patch panel. However, it is more difficult to see and to access the jacks on the side of the block, and potential wire management methods are poor. An addition, you must be certain that the assembly is data-grade, so that some twist is maintained between the jacks and the 66 block's clips. This arrangement must not be used for Category 4 or 5 operation, unless it is certified to those specifications.

Several methods may be used to mount the 66M in a permanent location. Most commonly, the block may be directly mounted on a backboard. The typical backboard is a 4×8 sheet of ¾ inch plywood, securely fastened to the wall and painted an appropriate color, such as battleship gray. The 66-block assembly is actually made up of two pieces, a front block that contains the contacts and molded side slots, and a stand-off bracket that mounts to the backboard. The bracket is normally mounted to the backboard before the 66M is attached. Two slotted, mounting holes are located on the upper left and the lower right corners of the bracket. The slotted arrangement allows the mounting screws to be preinstalled on the backboard before the bracket is mounted. This two-piece structure, bracket and block, allows easy alignment, quick mounting, and facilitates the use of preassembled mounting frames. After the bracket is mounted, the 66 block is set on top and secured by four clips that are part of the bracket. Connectorized 66 blocks have the connector mounted to the bracket, with the front block already attached, and must be installed as a unit.

In addition to direct mounting on a backboard, the 66M may be mounted in a preassembled mounting frame. A typical distribution frame is shown in Figure 8.5. The frames are mounted as a unit onto wooden backboards, although some of the frames may be mounted on standard 19 inch rails as well. These frames usually have preinstalled mounting brackets, into which the

Figure 8.5 A typical distribution frame for 66 blocks. (*Courtesy of The Siemon Company.*)

appropriate 66M block can be snapped. The frames may also have plastic stand-offs (sometimes called "mushrooms") and brackets for cable and crossconnect wire management. Frames are available in various sizes and can be used to implement full intermediate or main distribution frame (IDF or MDF) facilities.

Various wiring color codes for the 66 block are shown in Fig. 8.6. The cables are terminated on the 66 block according to these standard color codes. The 4 pair code may be used to implement a TIA 568-A wiring system.

The overall color pattern is laid out in five "groups" of five pairs (10 wires) each. Each group of five pairs has one primary color that is the same for all of the pairs. (Note that these are not "primary" colors, as taught in art class; rather, the term refers to the first color in a pair of wires.) The primary colors, in order, are white, red, black, yellow, and violet (sometimes called purple) and are commonly abbreviated W, R, BK, Y, V (or P), although other abbreviations are sometimes used for readability, such as WHT, RED, BLK, YEL, and VIO or PUR. The first wire group, for example, has a white wire in each pair, along with a second wire of a different color. The secondary wire colors are blue, orange, green, brown, and slate (sort of a silver/gray) and are abbreviated BL, O, G, BR, and S (or sometimes BLU, ORG, GRN, BRN, and SLT). Note that the TIA 568-A uses the shortened abbreviations (the first letter of each color where there is no ambiguity and a second letter where needed). For example, for a 4 pair cable, the color abbreviations are W (the primary color) and BL, O, G, and BR (the secondary colors).

Each wire in a pair bears a helical or round stripe that is the same color as its mate. The pair is referred to by its primary and secondary colors. Thus the first pair of the first group is the W/BL pair (said "white-blue") and consists of

Contact Row	Pair	25 Pair Colors	4 Pair Colors	3 Pair Colors	2 Pair Colors
1	1	White/Blue	White/Blue [1]	White/Blue [1]	White/Blue [1]
2		Blue/White	Blue/White	Blue/White	Blue/White
3	2	White/Orange	White/Orange	White/Orange	White/Orange
4		Orange/White	Orange/White	Orange/White	Orange/White
5	3	White/Green	White/Green	White/Green	White/Blue [2]
6		Green/White	Green/White	Green/White [2]	Blue/White
7	4	White/Brown	White/Brown	White/Blue	White/Orange
8		Brown/White	Brown/White [2]	Blue/White	Orange/White
9	5	White/Slate	White/Blue	White/Orange	White/Blue [3]
10		Slate/White	Blue/White [2]	Orange/White	Blue/White
11	6	Red/Blue	White/Orange	White/Green [3]	White/Orange
12		Blue/Red	Orange/White	Green/White	Orange/White
13	7	Red/Orange	White/Green	White/Blue	White/Blue [4]
14		Orange/Red	Green/White [3]	Blue/White [4]	Blue/White
15	8	Red/Green	White/Brown	White/Orange	White/Orange
16		Green/Red	Brown/White	Orange/White	Orange/White
17	9	Red/Brown	White/Blue	White/Green	White/Blue
18		Brown/Red	Blue/White [3]	Green/White	Blue/White
19	10	Red/Slate	White/Orange	White/Blue [5]	White/Orange
20		Slate/Red	Orange/White	Blue/White	Orange/White
21	11	Black/Blue	White/Green [4]	White/Orange	White/Blue
22		Blue/Black	Green/White	Orange/White	Blue/White
23	12	Black/Orange	White/Brown	White/Green [6]	White/Orange
24		Orange/Black	Brown/White	Green/White	Orange/White
25	13	Black/Green	White/Blue	White/Blue	White/Blue
26		Green/Black	Blue/White [4]	Blue/White [7]	Blue/White
27	14	Black/Brown	White/Orange [5]	White/Orange	White/Orange
28		Brown/Black	Orange/White	Orange/White	Orange/White
29	15	Black/Slate	White/Green	White/Green	White/Blue
30		Slate/Black	Green/White	Green/White [8]	Blue/White
31	16	Yellow/Blue	White/Brown	White/Blue	White/Orange
32		Blue/Yellow	Brown/White [6]	Blue/White	Orange/White
33	17	Yellow/Orange	White/Blue	White/Orange	White/Blue
34		Orange/Yellow	Blue/White	Orange/White	Blue/White
35	18	Yellow/Green	White/Orange	White/Green [9]	White/Orange
36		Green/Yellow [5]	Orange/White	Green/White	Orange/White
37	19	Yel/Brown	White/Green	White/Blue	White/Blue
38		Brown/Yellow	Green/White [7]	Blue/White [10]	Blue/White
39	20	Yellow/Slate	White/Brown	White/Orange	White/Orange
40		Slate/Yellow	Brown/White	Orange/White	Orange/White
41	21	Violet/Blue	White/Blue	White/Green	White/Blue
42		Blue/Violet	Blue/White	Green/White	Blue/White
43	22	Violet/Orange [6]	White/Orange	White/Blue [11]	White/Orange
44		Orange/Violet	Orange/White	Blue/White	Orange/White
45	23	Violet/Green	White/Green [8]	White/Orange	White/Blue
46		Green/Violet	Green/White	Orange/White	Blue/White
47	24	Violet/Brown	White/Brown	White/Green [12]	White/Orange
48		Brown/Violet	Brown/White	Green/White	Orange/White
49	25	Violet/Slate			
50		Slate/Violet			

Figure 8.6 Various wiring color codes for the 66 block. Patterns for the 110 block are the same.

a white wire with a blue stripe and a blue wire with a white stripe. The primary-colored wire is always punched down first (from top to bottom of the block). Thus the top row of the punchdown block will have a W/BL wire, the second row a BL/W, the third row a W/O, and so forth.

Four-pair wires are a special case because they only use four pairs from the first primary color group. Because of this, there is no confusion if the pairs are referred to by their secondary color only, such as "the blue pair" or "the green pair." Just make sure you punch down each pair's white wire first. TIA 568-A also allows an alternate color coding for patch cords. See Chapter 9 for more information.

Type 110 connecting blocks

Another style of cable termination block that is in wide use is the type 110 connecting block. The 110-style block is a relative newcomer, compared to the older 66 block. Even so, it has been in use for well over a decade. The 110 system is designed for higher wiring density and a better separation of "input" and "output" cables than the older system. The 110 system is also important because the 4 pair connecting block component is used for the insulation-displacement termination in many patch panel and outlet jack systems by various manufacturers.

The 110 system consists of two basic components, the 110 wiring block and the 110C connecting block. The 110 wiring block, shown in Fig. 8.7, is a molded plastic mounting block with "horizontal index strips" that organize and secure 25 pairs (50 wires) each. Blocks are available that accommodate 100 and 300 pairs. The 100 pair block has four horizontal index strips, and the 300 pair block has 12. The 110A wiring block has a 3.25 inch depth from the mounting plane and is used in normal applications. The empty space below the block makes feeder cable routing easy. A 110D block is available with a 1.4 inch depth for low-profile special applications. The 300 pair wiring block illustrates the potential wiring density of the 110 system. It accommodates all 300 pairs in a 10.75 inch square footprint. A 110T disconnect block is also available that provides the ability to disconnect a circuit for testing in either direction.

Space is available between the index strips to neatly organize incoming cables. In addition, the tip (secondary) colors are marked on the strips to help the installer place wires in the block. The wiring block does not make electrical contact with cable pairs, but merely secures them in place. Once the incoming cable wires are routed onto the wiring block, they are electrically terminated by the insertion of a 110C connecting block.

The 110C connecting block is a small, waferlike plastic housing containing metal contact clips at opposite edges, as shown in Fig. 8.8. The blocks are designed to snap onto the wiring block strips and connect to the wires that are held in place there. The back edge of the wafer contains quick clip IDC prongs that resemble the straight portion of the 66 block's clips. When the connecting block is pressed into place on the wiring block, the wires positioned on the

Figure 8.7 The 110 wiring block is a molded plastic mounting block with horizontal index strips. (*Courtesy of The Siemon Company.*)

wiring block are terminated, in mass. Each connecting block can terminate from three to five pairs (six to ten wires), depending upon the size of the block.

The front edge of the 110C connecting block is used to terminate crossconnect wires, or occasionally, other cables and adapters. It is colorcoded to assist in the placement of wires, which are terminated on the top of the block one at a time. The IDC clips in the 110 block are designed to slice through the insulation as the wires are punched onto the top of the block. The action mechanically holds the conductors in place and makes electrical contact between the crossconnect wires and the underlying station cable wires that were previously terminated at the back of the block.

The routing of wires in and around the wiring block is one of the advantages of the 110 system. Station cables are routed to the wiring block from the rear of the wiring block. They are terminated on the back edge of the connecting blocks and are out of the way during the crossconnect operation. Crossconnects are made with unjacketed paired wire that passes in the troughs between horizontal index strips to reach the front edge of the appropriate 110 block. The positions on the wiring block may be easily identified by labels that attach to the block, unlike 66 blocks that have only a very small plastic surface for circuit marking. The labels fit into clear plastic label holders that snap onto the wiring block between the rows of the horizontal indexes strips. The labels are

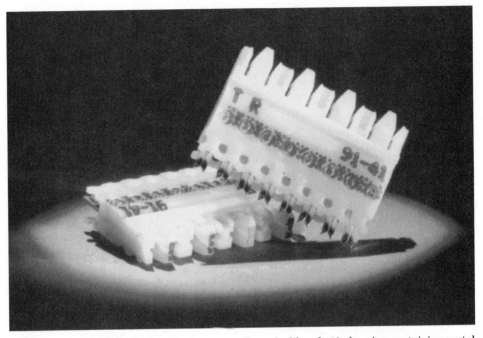

Figure 8.8 The 110C connecting block is a small, wafer-like plastic housing containing metal contact clips at opposite edges. (*Courtesy of The Siemon Company.*)

typically marked with thin vertical lines between station positions. Thus a label for 4 pair cable will have the lines every eight index positions. The labels do not interfere with the routing or tracing of crossconnect wires.

The 110 system is designed for AWG 22 to 26 plastic insulated wires. The 110 connectors are designed for *solid* wire only. Never use stranded-conductor wire with these connectors. The incoming station wires are inserted and trimmed using a 788-type impact tool, shown in Fig. 8.9. This tool can insert and trim up to five pairs (10 wires) at a time on the wiring block. It also seats the 110C connecting block onto a wiring block position. Crossconnect wires are terminated and trimmed one at a time with a punchdown tool that has a 110-type blade.

A 110 block system is used to terminate multipair station cables and allow the crossconnection to other punchdown locations. A typical LAN installation might have an appropriate number of wall-mounted or rack-mounted 110 blocks with crossconnect jumpers to a patch panel. Distribution frame assemblies are available that have preterminated 25 pair multiconductor cables connected to the back positions of the connecting blocks. The multiconductor cables terminate in 50-pin telco connectors that may provide mass connections to hub, telephone, or other equipment.

As we said before, each horizontal strip of the wiring block unit contains positions for 50 wires, or 25 pairs. This strip, in many ways, corresponds one side of the 66 block. Using the same math we did for 66 blocks, the 110 strip can accommodate six individual 4 pair station cables (or eight ea. 3 pair cables, or 12 ea. 2 pair cables). It can also terminate one 25 pair cable. As with the other type of block, 50 is an "odd" number, so the last two positions are not needed; only 48 positions are used. Contacts on adjacent 110 block strips may be connected together, one position at a time, by using the 110-type patch plug. The plug stretches vertically between two rows and connects the corresponding positions without having to use jumper wires. The plugs correspond to the bridging clips that are used to connect clips in adjacent columns on the 66 block.

In general, the 110 system is suitable for Category 5 use, as well as all lower categories. This is due the smaller geometry of the 110 clip and the ease of maintaining pair twist up to the very point of termination. Still, workmanship is an important consideration. Terminating wires on the 110 block is an individual operation and the individual doing the termination must be properly trained and sufficiently methodical. The station wires at the back of the connecting block are more difficult to inspect than those on the front, so care must be taken to maintain good quality terminations.

The 110 connecting block is also an important component in many outlet jacks and patch panels. It has the advantage of being extremely quick and

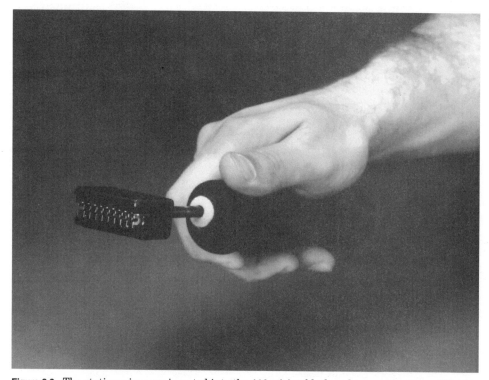

Figure 8.9 The station wires are inserted into the 110 wiring block and trimmed using a 788-type impact tool. Crossconnect wires are punched down using a 110 blade (see Fig. 8.4). (*Courtesy of The Siemon Company.*)

easy to use, in addition to providing a high quality connection. You can probably terminate a 110 block in half the time it takes with many custom IDC types, even though you must use a 110 punchdown tool. The comparison to conventional screw terminations is even better. The ready availability of 110 tools makes the block ideal for outlet and patch uses because it is likely to be in any installer's tool pouch. The quality of the connection is an issue for these jacks, just as it is for wiring closet terminations. The 110 block allows you to easily maintain the twist right up to the point of termination, ensuring a low-crosstalk and impedance-matched connection. The major disadvantage to the 110 block is that it positions the contacts in a line, and thus it might be a little wider than some of the custom IDC terminations. This restricts the side-to-side placement of jacks somewhat on the outlet plate. Some manufacturers of patch panels have dealt with this problem by simply placing the 110 strips in two conventional rows, with paired wires running to each jack.

Because of the design of the wiring blocks, the 110 system is not really suited for concurrent mounting of modular jacks, as is done with 66 blocks. However, models of patch panels with from 12 to 108 integral jacks are available. The jacks take up horizontal spaces, reducing the wiring density. It is difficult to offer a 50-pin telco jack mounting for the same reason. The 110 blocks tend to have cable-attached 50-pin jacks, rather than integral ones. Any 110 block that is thus wired to a 50-pin jack could use any of the fanout cables that were mentioned in conjunction with the 66 blocks, but you would lose much of the simplicity of this latter method. A better route would be to just use a patch panel with 110-type connections and either terminate directly to the patch panel or wire crossconnects over to the 110 wiring block.

The 110 system also offers a number of clip-on adapters that connect adapter jacks, cables, and other devices to the front of a 110C connecting block. Few, if any, of these are really practical for the orderly, structured wiring approach we have advocated for LAN wiring. In most cases, these adapters cannot be used for higher categories of operations, such as Category 5. However, they are handy for testing, as you can easily adapt cable scanners and other testers to the block wiring.

The design of 110 wiring blocks makes direct wall mounting, without a backboard, somewhat easier. This is due to the multiple rows of index strips that are an integral part of the wiring block. A 300 pair block could be mounted all at once, whereas the 66 block equivalent would require mounting at least six separate blocks. That would require a lot of wall anchors to mount to a hollow wall. However, we are not at all advocating abandoning the backboard. It is a part of the recommended standards for wiring closets, and is a very convenient way to provide a structurally sound mounting for many wiring components.

Relay rack and equipment mounting options are also available for 110 system wiring blocks. These methods are very good for large installations with the need for significant wire management.

The wiring color codes were covered in detail in the earlier section on 66 blocks and are identical for the 110 block system. The length of the horizontal index strips, on which the connecting blocks mount, are the same 50 wiring positions as the 66 block. However, the 110 system has the added advantage of placing color coding on some of the 110 components, to make placing the wires much easier. When you get to a very large installation, this can be quite an advantage, as it can also be used to visually inspect the terminations for correct routing.

Other Connecting Systems

We will mention two other connecting systems that are beginning to be used in more and more installations. One is the NORDX/CDT (formerly Nortel Northern Telecom) BIX system and the other is the Krone system. Both offer equivalent functionality to the 110 system and significant advantages over the older 66 system.

BIX connecting blocks

The NORDX/CDT BIX system is very similar in concept to the 110 system with a dual-sided 50-clip connector wafer that mounts horizontally in a mounting frame. Unlike the 110 system, the station or feeder cables are punched down directly into the back edge contacts of the connector. Some examples of BIX hardware are shown in Fig. 8.10. An array of BIX connectors, installed in a 50, 250, or 300 pair mount, forms a "module." Modules may be mounted directly on walls, backboards, or in racks with an optional rack-mount kit. Modules may also be combined into a mounting frame to form an intermediate distribution frame (IDF) or main distributing frame (MDF).

Crossconnects are made from the front of the BIX connector. Connectors are available premarked at either 4 or 5 pair intervals. The 4 pair version is used in most LAN wiring, while the 5 pair version is useful for termination of cables of 25 or more pairs.

Figure 8.10 A BIX connector. (*Courtesy of NORDX / CDT.*)

All connections are made with a BIX connecting tool, which is similar in function to the tools used for 66 and 110 blocks. Some standard impact tools can be equipped with BIX blades. Cables enter from the sides of the mount, either above or below the connector. The mounts include a provision for a marking strip (or designation strip), which mounts between a pair of BIX connectors. The connectors are used in pairs, with a marking strip in between. A common 300 pair mount, for example, contains a maximum of 12 connectors and six designation strips. Connector labels are available for the 4 pair connectors in colors to match the usage group designation in EIA/TIA 569. (See Fig. 11.2 in Chapter 11.)

Bridging clips are available that connect vertically between two adjacent connector positions. A special wiring fixture can also be used to assist in terminating cables. The fixture is snapped into a connector position and is moved to the next connector position to be terminated, in turn.

Modular jack assemblies that mount in the BIX mounts are also available in a variety of jack configurations. The assemblies occupy two connector slots in the mount. Category 3, 4, and 5 versions are offered.

The BIX-type connecting blocks are also used in modular jack outlets and on patch panels for wire termination. They are common options from several manufacturers, although all of the blocks come from NORDX/CDT. Distribution modules, jacks, and patches using BIX connecting hardware may be certified for Category 5, although some accessories and assemblies are certified to lower categories. The usual caveats apply to maintaining pair twist up to the point of termination, as with the other systems.

Krone connecting blocks

The Krone (pronounced krohn-a) connector system, shown in Fig. 8.11, features 8, 10, and 25 pair basic connector modules that can be mounted in several configurations. The 8 and 10 pair connectors can be installed in separate module mounts that each have a capacity of 20 connectors. These mounts can therefore terminate a total of either 160 or 200 pairs. The 8 pair connector is typically used to terminate two 4 pair station cables. The module mounts can be individually wall-mounted or can be installed in triples in a 19-inch wall or rack-mount frame.

For 25 pair applications, a 25 pair connector and mount is available. The 25 pair connector module is a feed-through design with a front and back piece, similar to the 110 system. As with the 8 and 10 pair connectors, the 25 pair connector is installed horizontally into its mounting bracket. A special assembly of two of the 25 pair connector modules is available with a mounting bracket that is the same size and footprint as the standard 66 block. These assemblies terminate 50 pairs, as do 66 blocks, and are ideal for replacement of the older blocks for higher performance applications. They also mount in distribution frames intended for 66 blocks. The availability of 66-type mounts means that versions of the block can be offered with integral 50-pin telco connectors, just as with the 66 blocks.

Figure 8.11 The Krone connector system features 8, 10, and 25 pair basic connector modules that can be mounted in several configurations. An 8 pair module is shown. (*Courtesy of KRONE, Inc.*)

Station and crossconnect wires are terminated on the bottom and on the top, respectively, of a Krone connector. The connector offers a unique "disconnect" contact as an option. Connectors with the disconnect feature can be temporarily opened for isolated testing to either leg of the circuit.

Krone connectors may also be found in outlet jacks and patch panels for wire termination. Many manufacturers offer the Krone wire termination system as an option. The connector hardware may be certified to Category 5, but again, the usual guidelines for wire pair termination must be observed.

Routing and Dressing Devices

Cabling, crossconnects, and patching in wiring closets should result in an installation that is neat and orderly. Unfortunately, all these wires and cords have a natural tendency to be very disorderly, it seems. To avoid the mess and the stress it brings, you should use the proper wire-management devices to put the wire in its place. Proper planning, layout, and dressing-in of the wire and patch cords can result in a very nice installation.

The process of properly routing and dressing your cable can keep your wiring closet much more "user friendly." If you are the user, you will really appreciate the neatness every time you need access to your cable system. If you are the installer, your customer will be much more pleased with your installation. The system of routing and dressing-in (putting in place) twisted pair wiring is a relatively simple. The principal is that all the wires and cables should be run along wiring channels or trays, secured in place with cable ties or other devices, make relatively square corners, and be out of sight as much as possible when securing the wire is not practical.

You can divide the wire management problems into two separate issues. One issue is how you should deal with relatively permanent wiring, such as horizontal station cables and crossconnects. The other issue is how you manage temporary wiring, such as patch cords (and equipment cords).

Station cable and crossconnects consist of solid core wires and cables. These wires are relatively easy to bend into position, wrap around standoffs or brackets, and secure with a few cable ties. The solid wire tends to bend into place and stay with a minimum of restraint.

The stranded-wire cable used in patch cords is not as well behaved. It is used specifically because it is so flexible, relative to solid-wire cable. These patch cords are truly a mess to deal with, leading some to try to avoid their use entirely and use semipermanent crossconnect wire instead. In LAN wiring, flexible equipment cords are often used between a patch panel and the LAN hub. Many hub ports must be connected with an 8-pin modular cord, unless they are equipped with mass-terminated connectors, such as the 50-pin telco connector. Thus, the use of stranded-wire cords may be unavoidable.

How you handle cable management will be a function of the size of your wiring closet and how it is laid out. If you have a small, wall-mounted wire termination system with a few wall-mounted hubs, you may be able to deal with the cables by securing the station wires with tie wraps and standoffs, running crossconnects (if any) around standoff posts or through cable rings, and letting the patch or equipment cords droop (neatly, we hope). On the other hand, if you have a large facility, or have equipment and wiring devices that are mounted in floor racks, you should use a system of cable trays, panels, raceways, and brackets that routes all types of cable and keeps them very neat. There is nothing more troublesome than having to pick your way through a curtain of patch cords to find a patch jack. There are much better ways to install your cable and cords, and we will cover some of them here.

We should caution you first that the use of tie wraps (cable ties) that are excessively tightened can cause performance problems on Category 5 wiring and should be avoided. However, relatively loose ties that do not distort the cable jacket should be all right. The effect is cumulative, so 20 tie wraps are worse than five. You can use a new type of wrap made with hook-and-loop mesh (Velcro-style) in lieu of traditional nylon tie wraps. This new wrap is wider, so it does not pinch the cable as badly. It is also easily removable so that new cables can be added without adding tie-wrap clutter.

Standoffs and distribution rings

The most basic accessories for wire management are plastic standoffs and plastic or metal distribution rings. Some examples of standoffs are shown in Fig. 8.12. The plastic standoffs (sometimes called *wire spools* or *mushrooms*)

Figure 8.12 A basic accessory for wire management is the plastic standoffs. (*Courtesy of The Siemon Company.*)

are designed to hold cables or crossconnect wires underneath the outer lip (which gives it the mushroom look). Wires are wrapped over a standoff, or down the side of a line of standoffs and routed to their destination. The wire is usually bent around the post very slightly to give the wire a "set" and hold it in place. Tie wraps may also be used to hold wires in place, subject to precautions applicable to Category 5 use. The standoffs are hollow and may be secured to the wiring board with a captive wood screw or machine screw. Standoffs are also supplied without a screw. These devices are very widely used in crossconnect fields, especially with the 66-style connecting block, which has little native cable management.

Metal distribution rings are often used on wooden backboards to secure station cables and crossconnects as they are run across the board. Cables may be quickly laid in, sorted, wrapped, and secured to the rings. The rings are available in several sizes, although the 2-, 4-, and 6-inch widths are the most common. The rings mount with two screws at either side. Some installers split the rings in two with a tubing cutter to provide an open half-ring for crossconnect wiring. This method allows more wires to be controlled than with the plastic standoffs and avoids the problem of threading the jumper wires through each closed ring as the wires are run from point to point.

The half-ring idea probably inspired a version of split plastic ring, also called a wire hanger or bracket, shown in Fig. 8.13. This bracket has a flat, solid back with side loops that almost meet at the top to form a ring. The small opening at

Figure 8.13 Split plastic rings are useful for managing wiring with frequent changes or additions. (*Courtesy of The Siemon Company.*)

the top allows a cable to be placed directly into the ring without threading. These brackets work well with stranded-wire cable or crossconnect wire, but the solid, plenum-style wire tends to work out of the bracket if not secured with a wrap.

Wire management panels

Wire management panels are available that mount horizontally across a relay rack or cabinet to offer orderly routing for patch cords and equipment cords. These panels prevent cable droop that would obscure equipment front panels and connections. The panels may consist of a series of split-ring loops similar to the wire hanger described above, or they may be a semienclosed slotted raceway with a removable cover. The styles with the cover offer the ultimate in out-of-sight cable management. They are great for patch cords, which present the greatest challenge to neatness. The covers are quite easy to remove, so you can get to the cords easily for new connections and moves. Covered cable raceways are also available for the vertical runs, although the vertical pathways are not as much of a problem.

With the increased concern over tie wraps and sharp bends when using Category 5 cabling, the gentle side of wire management really gets a boost. Now that many in the industry have heard of actual cable failures that were cured when the tight tie wraps were cut, you can bet that installers will pay more attention to cable management issues.

User Equipment Location

The proper location of user equipment, such as LAN hubs, is a key part of creating a successful wiring closet. While the wiring closet does not necessarily contain any equipment or hubs at all, at least in the grand scheme of structured wiring, your LAN wiring closet almost certainly will. These hubs and other equipment must be connected to the horizontal wiring for your workstations in order for your network to function.

The needs of LAN hubs and other active network components are a little different from the considerations for horizontal wiring terminations, crossconnects, and patches. Wiring and termination components simply need a fixed mounting location, wire management devices, and accessibility. The hubs and other equipment need power, ventilation, and connectivity to their respective station links. The wiring components tend to be fixed in place for the life of the installation. The equipment components, however, may need room to grow and to be reconfigured as network needs change and as technology changes.

The best location for the equipment is one that places the hubs and the wiring components as close together as practical, while maintaining a logical separation of the two functions. Remember, too, that some expandability of horizontal wiring connections should be allowed, and that correspondingly more room should be allowed for future changes in equipment technology.

In a small installation of less than 100 station terminations, you have a choice of wall or rack mounting for both the equipment and the cable terminations. There are advantages (and disadvantages) to both. Wall mounting will give you the most free and open space in the center of a small wiring closet. However, it limits access and cable routing options for the wiring components, and wall mounting of hubs and other equipment may be difficult. Rack mounting may be used both for wiring components and for hubs, but you will have added expense for the rack, cable trays, and other accessories.

In larger installations, serious consideration should be given to rack-mounting all components. This is particularly appropriate with the wide range of wire management devices for cable racks. You may face an interesting dilemma trying to decide in which racks to put the hubs. Should they be in the same racks, or should they be in adjacent racks? While there may be a certain logic to separating the wiring termination blocks and patches from the hubs, remember that you must generally run lots of patch cords between the two. With separate racks, all those patch cords would have to pass up through one rack and across and down into the other rack to make connections. Not only is such a long run probably unnecessary, but it may stretch you to the limit for the channel link. Remember that you are allowed only 10 m (33 ft) for all equipment cords in a channel, including both the user cords in the work area and the patch/equipment cords in the wiring closet. If you required a 5-m (16-ft) equipment cord to reach to the hub rack, you would only have a 5-m (16-ft) allowance in the work area. It is not unusual for larger work area offices to need 6- to 8-m (18- to 25-ft) user cords, so you would be better off using as short a wiring closet equipment cord as possible.

The easy way to keep patch and equipment cords short is to locate the horizontal wire terminations and the hubs in the same rack. The terminations might be on connecting blocks, with crossconnects to patch panels or directly to patch panels. You can use wire management panels to route the patch wires to the side rails and down to the hub. Most modern wire termination blocks feature a very high connection density, so they should not take up much rack space. Also, patch panels with 96 or more jacks consume very little vertical rack space. So it should be possible to place both equipment and terminations in the same rack, with plenty of room for future expansion of both. A popular alternative mounting scheme is to place the wiring terminations in wall mountings and place the LAN hubs in relay racks. This hybrid approach is serviceable, as long as the equipment cord runs are not too long from the wall to the racked hubs. In a variation to this approach, you could place the connecting block terminations on the wall, as before, and run crossconnect jumpers to patch panels in relay racks. Then, relatively short equipment cords could be connected between the patch panels and hubs in the same rack. Remember that the crossconnect jumpers count as part of the 10-m (33-ft) limit.

"To patch or not to patch, that is the question." You can get into a considerable debate with some wiring managers of large facilities about this very question. There is definitely a down side to the use of patch panels in a large facility. The problems with patch panels are that they are an added expense, provide an additional point of failure, may provide questionable long-term connections, and add lots of spaghetti-mess (from the myriad of patch cords that must be used). Their modular connectors may also add unnecessary crosstalk. If connections to the hubs can be made with crossconnect wire, you may be able to avoid the use of patch panels altogether. Simply terminate your station cables on the connecting block of your choice and run a crossconnect jumper over to a mass-termination connector for the hub. The larger card-cage hubs offer port connections on 50-pin telco connectors, among other options. The hub's 50-pin connectors can be extended to connecting blocks, such as the 110 system blocks, and the crossconnects can be made directly to the station punchdowns, without using patch panels.

The patchless approach has at least three disadvantages. First, troubleshooting is more difficult, as you cannot simply unplug a modular equipment cord and insert a tester. You must either disconnect the crossconnect wire and use a test jack adapter that fits onto the connecting block, or you must have used one of the types of connecting blocks with a built-in disconnect capability and still use a test adapter. Second, it is much more difficult, on average, to locate the connections for a particular station cable, since the circuit markings are imbedded in the high-density connecting block. Third, you may double the potential workmanship problems. For example, if you had directly terminated on a patch panel, that would be the only field termination that might have excessive untwist or other problems. The patch cords would hopefully be factory terminated, carefully inspected, and certified. With crossconnection, you add four points for potential workmanship problems: the connecting block, the two end connections of the crossconnect jumper, and the multicircuit termination leading to the hub.

Which is right for you? That may depend upon the size of your facility and the training of LAN maintenance personnel. The patchless approach should probably be avoided in wiring closets under 100 to 200 connections, unless you have trained people with the ability to do complex wire tracing and the workmanship skills to make high-quality connections.

Fiber Optic Termination

Fiber optic cables are treated a little differently in the wiring closet than conventional metallic cables. However, the concepts are the same. Basically, in the wiring closet, cables from the horizontal distribution to work areas are all brought together in some fixed, orderly arrangement for interconnection to network hubs and other equipment. Fiber optic cables may be used for the horizontal cabling, in place of traditional metallic wire. Fiber optic cables are also widely used for the backbone cabling that runs between wiring closets. Differences in the wiring closet terminations for fiber optic cables exist because the nature of the signals (light) and the fact that the transmission medium (glass fiber) requires special handling. These differences limit the number of ways the cables may be terminated, severely limit routing and handling considerations, and cause fundamental differences in the structure of patch panels.

In addition, fiber optic connections have what might be called "polarity." Because a single fiber sends its signal in only one direction, a fiber optic link always requires two fiber connections, one to transmit the signal in each direction. To operate properly, the transmitted signal from one end must connect to the receiver port at the other end, and vice-versa. This is called a crossover connection.

In contrast, metallic cable link standards use straight-through wiring for most connections. This is accomplished by defining two types of equipment interfaces, such as the DTE and DCE interfaces of RS-232 that are meant to connect "terminal" and "communication" devices. The same type of pattern exists in 10BaseT (adapter vs. hub) and AUI (repeater vs. transceiver). However, the fiber optic equipment interfaces have traditionally been identical, providing signal identification only by the label beside each of the two fiber optic connectors. That means that the user/installer must provide fiber cabling that accomplishes the crossover.

AB/BA polarity orientation

Fiber optic cabling standards, including 568SC and FDDI standards, specify a dual fiber connector that helps maintain polarity. The TIA 568-A method uses a polarized AB/BA orientation of the dual connector to incorporate polarity into the fiber connector. The method proposes to make the user connections foolproof, so that a user never has to worry about polarity. If the fiber cables are properly installed, the proper polarities at the user connections are assured. Unfortunately, that places all of the burden of maintaining proper polarity on the installer.

The polarity with the SC connection is achieved by using a nonsymmetrical dual-connector body and labeling each position either A or B, as shown in Fig. 8.14. The actual SC connector fiber termination was discussed in Chapter 7. The fibers of a given cable are not literally called "transmit" or "receive," but are numbered consecutively, paired, and alternately placed into either the A or B position at each cable end in such a way that the transmit connector on the equipment at one end is connected to the receive connector on the equipment at the other end. This is a variation of the old metallic cable proposal that would have made all equipment interfaces the same and all cables crossed over (transmit and receive reversed end-to-end). It is perhaps ironic that the metallic version was abandoned.

If you are accustomed to placing wire in connectors according to pin numbers and color codes, the fiber optic polarity method at first may seem confusing. Don't worry, the confusion will probably not go away, it will just become more tolerable. Figure 8.15 shows the interconnection of two duplex cables at an adapter interface. The term *adapter* is defined as a fiber optic coupling between two fiber connectors of the same type, that is, two SC connectors, in this case. (The term *hybrid adapter* is used to describe a coupling designed to mate to unlike-like fiber connectors, e.g., an ST and an SC.) Note that the adapter actually docs a polarity reversal by having opposite orientations between its two sides. The A connector position is always placed into the A adapter position, but that passes through to a B connector position, through the panel to the opposite side of the adapter.

The end-to-end connections of all fiber cables are also reversed. This includes the horizontal and backbone fiber cables as well as all patch cords and user equipment cords. The pattern of reversals should be standardized in an installation, but it is up to the installer to maintain this discipline. One way to accomplish this is to designate a correspondence between the fiber buffer color and a fiber number. Then terminate all odd numbered fibers as A

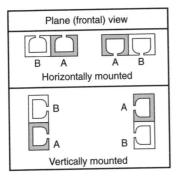

Figure 8.14 The polarity with the SC connection is achieved by using a nonsymmetrical dual-connector body and labeling each position either A or B. The two orientations are called AB and BA, reading either left to right or down.

Figure 8.15 The interconnection of two duplex cables at an "adapter" interface.

fibers and even numbered fibers as B fibers in the wiring closet. Do the opposite at the station outlet. For example, in a two-fiber cable, you would designate fiber 1 as an A connection in the wiring closet and as a B connection at the outlet. (Of course, the adapters at each end reverse the A and B, but that takes care of itself.) Be very consistent in your fiber polarity designations at every fiber termination. The scheme for backbone interconnections should be determined before any work begins and should be consistently applied. Remember that all fiber cable connections must reverse end-to-end for this scheme to work.

A couple of typical fiber optic connection patterns are shown in Figure 8.16. Note that the scheme must have an odd number of reversals to work, but that this is automatically accomplished by the fact that the coupling devices at the end of each run do reversals, too. Because cables always have the same type of male connectors, each patch cord, horizontal cable, or backbone cable always provides one reversal. The adapter couplings at either end provide two additional reversals, restoring the signal to the original polarity. All that you (or your installer) must remember is where to clip each connectorized fiber into its duplex housing to provide the proper AB/BA orientation.

If you do fiber to the desktop, you will want to install a fiber patch panel in your wiring closet. This panel will provide a point of termination for each station drop and will also provide labeling and identification for the horizontal run. It will also provide a handy point for testing and troubleshooting of fiber runs. An all-fiber installation will usually involve locating network hubs with fiber optic interfaces in the wiring closet. You will connect a duplex fiber optic patch cord between the station location on the patch panel and the hub port. You may have to use an adapter cable or hybrid adapter if your equipment does not have the new SC connectors. All of the standard cable management

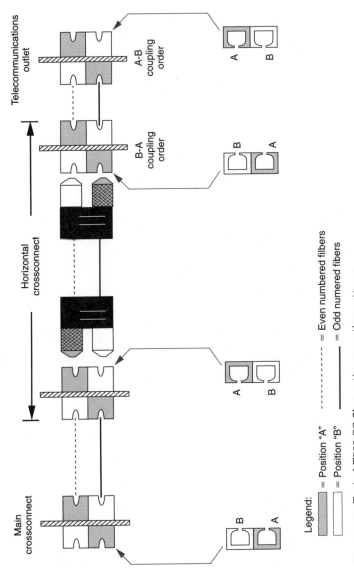

Figure 8.16 Typical T568-SC fiber optic connection patterns.

trays and panels may be used with fiber patch cords, in general, although you should pay particular attention that the cords are not bent sharply or otherwise stressed. The tiny glass fibers in these cables are subject to stress fractures when excessively bent, and may cause signal loss that prevents a good connection. A sharp bend may actually break the fiber entirely. Suspected cords should be tested and scrapped if bad.

You will want to carefully plan the location of your fiber optic terminations in the wiring closet. Because of the nature of fiber optic cable, it is difficult to reterminate when you move the termination location. Fiber optic cable is often terminated in a special fixture in the wiring closet. Fiber optic termination fixtures, as shown in Fig. 8.17, often have space for extra fiber to be wound. However, the extra fiber is really intended to allow some slack in case a damaged connection needs to be replaced, and to assist in the termination process by allowing greater access to the fiber end. Usually, the amount of extra cable is only 1 m or so, and may have already had the outer jacket removed to expose the fibers, so it would not be useful in a relocation.

Figure 8.17 Fiber optic termination fixtures often have space for extra fiber to be wound. This multimedia example also has copper wire jacks. (*Courtesy of MOD-TAP.*)

The use of a termination fixture, sometimes called a break-out box or fan-out enclosure, makes free-standing racks the most convenient location for large numbers of fiber optic terminations. This certainly is the case if you are using fiber for the horizontal connections. The fiber is typically routed across cable ladders or closed bottom trays to each rack, and then down along the inside or back of a rail to the termination location. Fiber optic cables may be carried in a brightly colored, corrugated plastic "innerduct" for additional protection. The use of innerduct is mandatory if you use unjacketed fibers or highly flexible fiber cables, such as "zip cord" cable. The innerduct has an additional advantage because of its fairly rigid walls. It can be tie-wrapped in place with little chance of damage to the interior fiber cables. Also, it bends easily, but keeps a fairly large bend radius because of its construction. Fiber cables can often be placed in existing innerduct much as you would in a rigid conduit.

Caution is the watch work when you are working with fiber optic cable, whether terminated or not. Terminated fibers may be connected to equipment that emits hazardous levels of laser light, especially if you are using single mode fiber. Appropriate eye protection should be worn and appropriate precautions taken when you are dealing with laser emitters that transmit above Hazard Level 1. ANSI Z136.2 contains more information on these precautions. Even nonlaser light may be a hazard if held close to the eye. A good rule is to **never look into a fiber**. The fiber wavelengths are in the invisible infrared spectrum anyhow, so looking only subjects you to a hazard, as you cannot see the light. The hazard exists even with unterminated fiber ends, so treat them as you would terminated fiber. Although there are visible light testers that are sometimes used during troubleshooting, you would be safer when using these if you simply point the connector or fiber at your hand and look for a dot of light, rather than trying to look directly at the fiber. Bad habits are hard to break, so don't start one!

The tiny glass slivers that result from cutting, cleaving, and sometimes polishing fibers are also an eye hazard. Always wear safety glasses when terminating fiber. You also don't want to inhale any of the fiber dust, but you should be able to avoid it when field terminating fiber cables under normal circumstances.

Marking

As we discussed for station outlets, proper identification of wiring closet terminations is crucial to a high quality, serviceable wiring installation. This includes a consistent method of cable and termination numbering.

Each and every cable, termination fixture, patch panel, and crossconnect should be clearly identified. EIA/TIA 606 provides a method of marking just about everything in the telecommunications wiring world, with wiring terminations being no exception. The guidelines call for each item to be marked with a unique identification that is clearly visible. That generally means you should use ⅜-inch high lettering and designate each cable termination and patch position in some orderly fashion, such as with the wiring closet number, cable number, position, etc.

Some connecting blocks make marking difficult, while others seem to have marking panels as part of the design. Contrasting examples are 66 blocks and 110 blocks. Some older patch panels leave little room for markings, other than the small factory marked jack number. In a very small installation, this does not cause too much difficulty, but larger installations require more detailed, explanatory markings.

You also may choose to use color to help designate wire termination areas. The official colors for grouping different types of cable terminations are given in Chapter 11. The colors are defined in *EIA/TIA 569, Commercial Building Standard for Telecomunications Pathways and Spaces.*

Workmanship

One of the keys to a successful LAN wiring installation is good workmanship. You can use all the finest components, the most expensive mountings, and the highest ratings, and still have a poor installation due to improper installation techniques. The wiring closet is as much in need of good installation practice as any other area. In some ways, there are more places in a typical wiring closet to do things that might adversely affect your cable link performance than elsewhere. For example, you may make one station cable termination, two crossconnects, and one patch termination all in the same cable link. Add to that the patch cords and equipment cords, and there is a lot of room for error.

The first workmanship guideline is to properly route the cables. You should minimize the sharp bends, avoid tight tie-wraps, avoid proximity to magnetic fields, and limit exposure to mechanical damage. You should ensure that the pairs are kept twisted as close to the point of electrical termination as possible. See Chapter 7 for an example of the effect of slightly untwisted wire.

The standards for preserving wire pair twists are defined in TIA 568-A. For Category 5 terminations, the maximum amount of untwist is 13 mm (0.5 in). For Category 4, it is 25 mm (1.0 in). The twist must be preserved within these limits at each point of termination to connecting hardware, including connecting blocks. The practice is not optional, it is mandatory. There is no limit specified for Category 3, but good workmanship practices should maintain the twist to within 38 mm (1.5 in). Even Category 3 links can suffer from the increased crosstalk and impedance mismatch problems of excessive untwisted wire.

The cable jacket should only be removed to the length necessary to terminate the pairs. This means that you do not strip back the jacket 4 or 5 inches, but only about 25 mm (1 inch) more than is needed for the termination of a single pair.

When punching down wires in a 66-type block, there is a possibility that another adjacent wire may accidentally get cut, along with the wire you intend to terminate and trim. The punchdown tools usually have one bright yellow side with the word "CUT." If you wrap the wire in from the top of the contact, the cutting edge of the blade should trim excess wire on the bottom. To do this, the CUT side is down. If you see the word CUT, turn the tool around so you don't cut the wire in the wrong place. If you are using the minimum-untwist technique for Category 5 we described earlier (see Type 66 connecting blocks at the beginning of this chapter), you must use the tool in both orientations, as one wire of a pair will enter its contact from the bottom and the other from the top. This takes a little extra thinking and dexterity to master, but can done quickly once you get used to it.

Do not leave the cut-off excess wire in between the contact rows of the block. You may accidentally force it into a contact slot or cause a short when you terminate another wire. Neglected trimmings will also make wire verification and tracing more difficult.

The requirements of TIA 568-A dictate a minimum bending radius of 4 times the cable diameter for certain categories of operation. The time-honored technique of pulling the station cable very tight and bending it back to the block before terminating the wires will have to be abandoned. You should leave the cable somewhat loose if you are trying to maintain this minimum bend radius. Many newer types of connecting blocks have wire management pathways, clips, ties, and other devices to prevent sharp bends in the cables. If your termination blocks do not have these accessories, you will need to use good habits when you terminate cables, in order to avoid problems. Remember that these guidelines also apply to crossconnect cable and patch cords. Pair twist must be maintained and the bend radius guidelines observed to maintain maximum rated cable link performance. A summary of workmanship and installation rules is shown in Table 8.1.

TABLE 8.1 Typical Workmanship and Installation Rules in the Standard

Document	Reference	Rule	Measure	Applicable to:
TIA 568-A	10.6.1	Terminate w/connecting hardware of same or higher category		Horizontal cables
	10.6.3.1*	Minimum amount of untwist at termination, for 100-ohm cable:	13 mm (0.5 in)	Cat 5
		Minimum amount of untwist at termination, for 100-ohm cable:	25 mm (1.0 in)	Cat 4
	10.6.3.2	Maximum pulling tension	110 N (25 lbf)	4-pr
		Minimum bend radius	>4x the cable diameter	4-pr
			>10x the cable diameter	Multipair

*Mandatory requirement

9

Patch Panels
and Interconnection

Patch panels, patch cords, and crossconnect jumpers form a vital part of LAN wiring. In this chapter, we will describe the wiring, cables, and devices used for crossconnects, patch cords, and circuit patching in cabling systems designed for LAN use. In the last chapter, we covered the actual termination of the station cables into punchdown blocks in the wiring closet. We did not yet cover the crossconnect wiring or jumper cables. In addition, although we also included the wiring method of terminating cables directly into patch panels, we have waited until now to describe the patches themselves.

There is an ongoing debate about the alternatives of patch panels and crossconnects. Figure 9.1 shows the two alternatives to cable termination and the interconnection to equipment in the wiring closet.

The first alternative shown in the figure copies the crossconnect methodology used for traditional telephone equipment room wiring. In this method, horizontal (station) cable is terminated on a connecting block, such as a 110 or 66 block. Then crossconnect wires are run from each station position on the horizontal connecting blocks to a second field of connecting blocks, which is a point of termination for multipair cables from the common equipment such as a telephone switch or PBX. This common equipment connection in a telephone system corresponds to the network hub connection in a local area network.

The common equipment connections and the station cable connections are rather like the input/output of the crossconnect field. Both types of cables are hard wired to the connecting blocks and then crossconnected with unjacketed wire pairs called *jumpers*. To make a connection across the field, a crossconnect jumper (consisting of one to four twisted pairs) is punched down on the appropriate contacts of a common-equipment connecting block, routed across the frame, and the other end punched down on the contacts for the station cable. Changes require you to rip out a punched-down crossconnect jumper and install a jumper to the new location.

Figure 9.1 Two alternatives to cable termination and interconnection in the wiring closet. *(a)* Traditional punch down and crossconnect. *(b)* Direct patch-panel termination.

Connections in the punchdowns are theoretically gas-tight, so corrosion and signal deterioration over the long term are rare. This method works fine for LAN wiring, although you must pay attention to the issues of pair twist, minimum bend radius, and other cable-care considerations when wiring LAN crossconnects, just as you did with the installation of the station cable. This first method is pure crossconnect wiring, without the use of any patch panels.

With the second alternative, the horizontal cables are directly terminated into modular patch panels (unless you use a hybrid method of termination, described later). Patch cords are connected between the patch panel jacks and the equipment hubs, and connections may be reconfigured at will. This method is very convenient for smaller installations that use hubs (or other equipment) with discrete modular jacks. You must use some sort of modular connection to these hubs, and the patch allows the use of individual modular cords, and avoids the use of multipair fanout cables and connectorized punchdown blocks. It is also handy for companies that frequently move connections or add stations to the network.

Testing a patch panel installation is very easy. You simply disconnect a patch cord and connect your tester. The patch jacks are also more spread out than the cable termination areas on crossconnects and this may make it easier to provide proper circuit marking. (If you have ever tried to mark on the little tabs of a 66 block, you can appreciate this a lot.) Newer designs for patch panel terminations use IDC connections that are similar to those on traditional connecting blocks or outlet jacks. These designs help preserve the pair twist on terminated cables and use internal jack wiring that is designed to meet the new categories of performance.

Direct patch panel termination of station cables, however, limits the flexibility of a terminated cable, because the station cables pairs are available only on the corresponding patch jack. To gain access to unused cable pairs, an external adapter or splitter must be used. Because of this limitation and other factors, there are differing opinions on whether you should use patch panels or crossconnect fields.

Obviously, there are valid arguments on each side of this question. On one hand, when you terminate your horizontal cables directly onto a patch panel, you avoid the muss and fuss of at least three terminations (the horizontal punchdown and the two crossconnect terminations). Using the direct patch approach also saves space in the wiring closet, saves installation labor, and provides for quick reconfiguration of your network connections. On the other hand, you still have all the patch cables to deal with, and each of your station cables is forever terminated in a single 8-pin jack. You have no easy way to grab a pair or two to use for another purpose, such as a modem line or a second network connection. In addition, the patch cords are normally available only in standard length increments (unless you make your own), and thus rarely fit a connection exactly.

On some points, the two opposing cable termination techniques each have disadvantages that compensate for their advantages. For example, the exclusive use of crossconnect fields eliminates the crosstalk and contact resistance problems of patch panel jacks, but it compensates by introducing two more terminations (each end of the crossconnect wire) where the finest workmanship may still introduce untwist to the link (thus causing more crosstalk, etc.). Also, the ease of moving connections on the patch belies the fact that an average mature installation may actually need only 2-5% of its connections moved in a year.

Although we have mentioned two termination techniques—direct patch termination and crossconnect field termination—a third hybrid method uses a combination of both. In this method, the horizontal cable is terminated in a connecting block (such as a 110 block). Then a crossconnect is made between the connecting block and a patch panel. The patch panel may allow for directly connecting the crossconnect wire, or it may be connectorized, requiring multipair cables that terminate in a second connecting block. The patch cords are connected from the patch panel to the hub equipment. This hybrid method offers wiring flexibility between the station cable and the patch: a cable may easily be reconfigured, shared, or moved to another patch position. Equipment connections may be made quickly or reconfigured at will, due to the use of patch cords. However, the patch jacks and cords are still there, with all their other potential disadvantages.

It's almost impossible to take sides in this argument, as there are valid reasons behind both viewpoints. However, you should know that there is a definite choice to be made and to have an understanding of the issues surrounding that choice.

The next few sections will describe crossconnect wiring (the connecting blocks were covered in Chapter 8), multipair jumper cables, patch panels, and patch cords.

Crossconnects

If your station cables are terminated on punchdown blocks, you must connect each cable to its ultimate destination by using a crossconnect jumper. Crossconnect jumpers consist of un-jacketed, twisted pairs that are loosely spiraled around each other. The pairs are usually heat treated to make wires of each pair adhere slightly together. Crossconnect wire, as with station cable, is made from solid copper wire with plastic insulation. Unlike station cable, no covering jacket is used. The overall construction of each pair is similar to pairs used in jackcted cables, although the insulation may be a little stiffer to provide additional resistance to abrasion and cuts.

Crossconnect wire (it is usually not called cable) is available in 1, 2, 3, and 4 pair configurations. The pairs are usually color coded in the same manner as regular jacketed cables. The complete color code for these cables is shown in Chapter 3, but the short version is blue, orange, green, and brown pairs, each paired with a white wire. If you have only one pair, the blue pair (blue/white marking) is usually used. However, single-pair cable is also available in a blue/yellow combination, as that is often used for telco central-office lines, as well as other uses. If you are doing LAN wiring, you will want to use 2 pair or 4 pair crossconnect wire that is color coded the same as the station cable. To implement the full TIA 568, a 4 pair wiring system will requires the 4 pair cross connect wire, unless you make two runs of two pairs for each crossconnect. That standard suggests that crossconnect wire consist of pairs with one wire with white insulation and another with a distinct color such as red or blue. The normal color code scheme complies with that recommendation.

Wire size and type

Crossconnect wire is available in various wire gauges including 22 AWG to 26 AWG, but the most commonly used gauge is 24 AWG, as with jacketed station cables. The wire guidelines in TIA 568-A state that standard-compliant UTP horizontal cable should use 24 gauge wire, but that 22 gauge wire that meets the transmission requirements of the standard may also be used. It is vague about the wire size of crossconnect wire, requiring only that the wire meet the same transmission characteristics as regular horizontal UTP cable.

Crossconnect wire is normally used in wiring closets, so it does not need to be plenum rated. The normal insulation for crossconnect wire is PVC. Some local codes may require riser or plenum rating for these locations.

Wire category

The performance categories for crossconnect wire are the same as for horizontal cable. The proper category of wire should always be used when making your crossconnects. This means that a Category 5 installation will require Category 5 crossconnect wire and that lower category installations will require wire that meets at least the specified link category. You may always use a higher category rating, as with any cable or components. For example, you may use Category 4 or 5 wire in a Category 3 installation.

Because there is no jacket to show the wire ratings, the category of installed crossconnect wire is difficult to determine, unless it is Category 5 wire. Category 5 crossconnect wire has an extreme amount of twist to each pair, so it is easily recognized, because there is no plastic jacket to hide the twists. Category 3 wire is harder to distinguish from Category 2 or uncategorized wire, since the appearance may be similar. You may need to keep track of all the crossconnect wire used in your facility. The category, if any, should be clearly marked on the wire spool. If you are in an environment where you have total control over all the wire and cable, you may want to use Category 5, the highest category of crossconnect wire available, for all your crossconnects. Keep in mind, however, that the tight twists also make this wire a little harder to terminate, particularly in 66 blocks.

There are subtle differences between category ratings from different manufacturers. As is the case with jacketed pairs, consistency in the geometry of the pair has a big influence on the transmission characteristics of crossconnect wire. This means that pairs manufactured to very tight specifications can produce a superior jumper that will have a very low return loss, indicating very consistent impedance. This carefully controlled manufacturing also achieves better performance in some of the other transmission parameters. You can indeed buy better wire and cable, but you will probably pay more for the privilege.

Routing and marking

Crossconnect wires are routed between two termination points using the standoffs, brackets, and other wire management devices that were described in Chapter 8. Because of the lack of a jacket, crossconnect wiring may require a little extra care in handling to maintain performance characteristics. Some of the standard practices that are used in running telephone-grade crossconnects are not at all appropriate for the higher performance categories of LAN wiring. For example, it is common practice to sharply bend jumpers when they are wrapped around wire management brackets. These sharp bends may be less than the minimum bend radius recommended in cabling standards such as TIA 568-A. This standard calls for a bend radius to be not less than four times the diameter of the cable. Crossconnect wire has no jacket, so its overall diameter is not as easily determined, but it is probably less than jacketed cable. It is certainly easier to bend, as the jacket imparts additional stiffness to regular cables.

Connections on wooden backboards are fairly easy to route. If you need additional routing paths, you simply add the appropriate brackets or standoffs to the board. Wire is routed around or through the devices from start to end of a jumper. Distant blocks or patches just require longer crossconnect jumpers. If you use connecting blocks or patch panels that are in separate relay racks, as may be the case with large installations, you will probably need to route the wire up the side rail of the rack, across a wiring tray to the other rack, and down the other side rail. Wire management brackets, hangers, raceways, and trays are essential for proper protection and containment of the wire.

Crossconnects must be marked just as with all other wiring components, pathways, and cables. Obviously, the crossconnect wire itself is difficult to mark, so you will have to ensure that the points of termination of the crossconnect are marked clearly to indicate the identification of the crossconnect. The particular type of connecting block that you use will make a big difference in your ability to provide marking for each connection. Ideally, you would mark the termination in such a way that identifies both the point of termination and the associated connection at the other end of the crossconnect. In some cases, it may not be practical to mark anything other than the termination ID on the block. Some connecting blocks have additional labels, clips, covers, and brackets that can contain additional identification. However, some of these items have to be removed to terminate wires or to test connections, and thus they are easily lost.

You should keep comprehensive paper (or database) records of your interconnections to compensate for the shortcomings of connector block and jumper marking. The standards do allow placing a listing of circuit identifications adjacent to a point of termination, if the design of the devices makes direct marking impractical. Keep your records up to date, including all adds and moves, so that you can easily find the needed circuit links if you need to do troubleshooting or changes.

Workmanship

In addition to the routing and cable handling issues mentioned in the previous section, several other installation practices need to be observed when using crossconnect wiring for LAN cabling. Preserving pair twist is probably the most important single requirement to maintain the transmission characteristics of a link. Pair twist integrity becomes more critical with increasing link performance category. The very low-speed links, such as LocalTalk and Arcnet, will operate on very slightly twisted pairs. The moderate speeds of 10BaseT and 4 Mbps Token-Ring are not very particular about maintaining cable twist, as long as the amount of untwisted wire is minimal and the link does not push the distance limit. However, at the higher wire speeds of 16 Mbps Token-Ring, 100VG, 100BaseT, and ATM, the preservation of pair twist becomes increasingly more critical.

The effect of adding connecting blocks necessarily adds some amount of untwisted conductor paths to any cable run. However, you can certainly minimize the link degradation by maintaining the natural twist of wire pairs as close to the actual point of termination as possible. The maximum amount of untwist allowed by TIA 568-A is 25 mm (1.0 in) for Category 4 terminations and only 13 mm (0.5 in) for Category 5 terminations. No guidelines are given for Category 3 wiring, but many authorities recommend no more than 38 mm (1.5 in) of untwist. That should certainly be easy to accomplish. If you use even less untwisted wire at any termination, you will simply have a higher quality connection. The effects of untwist are somewhat cumulative, so neatness counts at every termination point in a link. You will find that some types of connecting blocks are more difficult to terminate with minimum twist than others. For this reason, many authorities do not recommend the use of traditional 66 blocks, even for Category 3 operation. However, some manufacturers do offer 66 blocks rated at Category 5 that should operate at the proscribed level of performance if the proper termination procedures are used. Chapter 8 describes a technique for minimizing the amount of untwisted wire when using a 66 block.

Many multipair cables are manufactured with a carefully controlled variation in twist pitch and pair spiraling in order to meet tough performance parameters such as NEXT (near-end crosstalk) and SRL (structural return loss). These parameters are maintained, at least in part, by the cable jacket that holds the pairs in their relative places. Crossconnect wire has no jacket, which means that you may have more difficulty keeping the pairs in approximately the same geometry, but you should try to do so if possible. It may be beneficial to use crossconnect wire that already has the number of pairs spiraled together that you need for the jumper connection, rather than making the connection with two or more individual jumper pairs. Using jumpers with the proper number of pairs will also make wire tracing and connection easier.

Remember to remove any unwanted jumpers completely. Nothing is more frustrating than having to fight through abandoned jumpers to trace down a connection or to make a new crossconnect. The crossconnect field should always be neat and orderly. The jumpers should be placed in wire management brackets or other devices and loosely bundled. (Hook and loop straps are good for this.) Never use tight tie wraps to bind jumpers together. The tie wraps distort the pair geometry and can produce the same performance degradation as with regular jacketed cables.

Punchdown (connecting) blocks are intended to have only one wire connection per contact. That means you should never punch down a second crossconnect wire in a slot where another wire is already punched. The integrity of the insulation displacement method used in these contacts depends upon the tight positioning of the contact jaws. The jaws are opened somewhat during the termination of the first wire, and a second wire will often not make good electrical or mechanical connection with the contact. The traditional 66 block is the most prone to this type of wiring error, because incoming and outgoing wires of a connection are lapped over each other as they are introduced to the block for termination. The crossconnect wires should be placed on the second (inner) set of contacts on this type of block. The 110 and similar types of connecting blocks are less subject to this type of error, because they physically separate incoming and outgoing wires on opposite sides of the contact block.

Patch Panels

Patch panels are used to provide flexible connections between horizontal station cables and the equipment ports in the wiring closet. Patch panels have modular jacks that are connected so that they correspond to work area outlet jacks or to equipment ports. An example of patch panels is shown in Fig. 9.2. The ports on LAN equipment, such as hubs, often use 8-pin modular jacks, so it is easy to connect a port to a patch panel by using a modular equipment cord (or patch cord). This provides sort of a natural LAN connecting environment, where all connections that the user makes involve familiar modular jacks and cords.

Patch panels are easy to install and are a little less confusing to the user than punchdowns, because each jack on the horizontal wiring patch corresponds to an identical jack at the workstation outlet. Patches also provide easy access, quick connections, and clear circuit marking (at least until they become covered with patch cords). Another aspect often overlooked is that adds,

Figure 9.2   Typical patch panels. (*Courtesy of The Siemon Company.*)

moves, and tests are all accomplished without using any tools. If appropriate wire management devices are used, the patched network can really be kept quite neat.

Specifications

A patch panel intended for LAN wiring use has an array of modular jacks that are mounted in a metal panel. The panel is generally wide enough to mount directly in a 19-inch equipment rack, while its height varies with the number of jacks on the panel. Panels may be of any size, with multiples of 12 or 24 jacks being common. Patch jacks often correspond to punchdown block connections, which dictate the multiple of 12 convention. For example, a group of 12 4 pair circuits can be accommodated by both sides of a 66 block (50 pairs), whereas a group of 24 4 pair circuits can be accommodated by a 100 pair 110 wiring block. Additionally, the number of 24 is convenient to use, because a row of 24 modular jacks fits nicely across a 19-inch panel width.

The connectors on a LAN patch panel are standard 8-pin modular jacks, wired either to T568A or B. The connections for station cables at the rear of the panel are usually color coded by pair number, which corresponds to the wiring pattern at the jack. It is possible to use other wiring patterns for special applications, but that isn't covered here. The connectors are spaced according to the manufacturer's preference, and you may find individually spaced connectors or connectors that are placed in groups in four or six adjacent jacks. The connectors are usually numbered according to their position on the panel. Optional circuit marking is described later in this section.

If you still have questions on whether to use the T568A or the T568B wiring pattern, see the discussion in Chapter 7 under "Modular jacks and standard jack pin-outs." For station cable wiring, it basically doesn't matter which you use, as long as the outlet jacks use the same pattern. Patch panels that use color-coded wire terminations are prewired to support either the T568A or the B pattern, so make certain which you order. You could use the opposite pattern for your needs, but you would have to reverse the orange and green pair (Pairs 2 and 3) and be forever confused when it came time to add new cables.

Standard patch panels will generally use some type of insulation-displacement connection (IDC) at the rear of the panel to terminate station cables (or other cables, such as those from nonmodular hubs and other equipment). The two basic design variations for patch panels either use groups of connectors that have a common termination block, or use individual connectors that have their own terminations at the rear of each connector. Another type of panel uses multicircuit connectors and is described in the next section.

The terminations for patch panels are identical in function (and often in appearance) to the terminations used in workstation outlet jacks or in connecting blocks. Variations in panels are available that use individual IDC connections, 110C blocks, 66 blocks, BIX blocks, and Krone blocks. Each method has the advantages and disadvantages that were described in Chapters 7 and 8, and you should choose according to your own preferences. It should be pointed out that the use of standard 66 blocks is not generally recommended for LAN wiring, particularly above Category 3. The use of newer Category 5 rated 66 blocks is still somewhat controversial, but if the manufacturer has UL-verified performance, you can probably use the blocks in this application.

Patch panels are connecting hardware, as far as the standards are concerned, and must be performance rated according to the category of operation you intend. Older-style panels paid little or no attention to preserving the pairing and twist of wires between the termination points and the modular jacks. You could get a nice glitch on a tester when you looked at an older patch panel in a LAN link. Fortunately, most modern panels are rated for Category 3, 4, or 5 operation. The TIA 568-A standard, as well as others, requires that all connecting hardware be marked as to its category of performance. The markings may either be the words "Category n" or "Cat n," where n is the category number 3, 4, or 5, or they may be a large capital letter "C" with the category number enclosed. If you do not see the marking, you should consider the panel unrated, and not suitable for LAN wiring use. Some manufacturers are providing a small category rating note along with the panel, certifying the category of performance. The only problem with this method is that it may be impossible to determine the rating of the panel after installation. You could provide the post installation assurance by affixing the notice to the rear of the panel, if a clear area is available. Still, clear factory marking of the category is the best bet.

The workmanship issues discussed in the previous sections on crossconnects and punchdown blocks certainly apply to patch panels. You must ensure that the cables to the back of the patch panel are properly routed and terminated. The termination procedures are the same as for the corresponding type of connection at a workstation outlet or punchdown block. The termination techniques and procedures to minimize untwist, kinks, and bends are described in Chapters 7 and 8.

Many of the potential problems that you will have with patch panels involve the routing and dressing (putting in place) of terminated cables and connected patch cords. The wide selection of wire management devices can help get the myriad of wires under control and out of the way. Some patch panels come with ingenious methods of cable routing and strain relief for the station cable terminations at the rear of the panel. The use of external wire management trays, such as the slotted raceways from several vendors, can eliminate the need for individual wire support and binding. Similar slotted raceways are available for management of the patch cords on the front of the panel. If the panel is not too many rows high, the patch cables can be routed vertically from the jack to the nearest cable raceway and tucked out of sight. From two to four rows of patch jacks can be accommodated easily, representing 24 to 48 patch positions on an average panel. At the side rails, vertical raceways can be provided to route the cords to patch or equipment positions on other levels of a rack. Cable trays and ladders (open bottom trays with cross pieces between the side runners) can be used to route the cords from rack to rack across the top of the racks. Computer rooms with raised floor panels can route the cables beneath the floor, but the above-rack method is much easier to maintain and can be made to look as neat.

Connectorized patch panels

A type of patch panel is available with preconnectorized jack connections, rather than terminations for individual cables. This type of patch is used extensively with the connectorized punchdown blocks, as described earlier, and it may be useful for the multi-port equipment connections of larger hubs. The connectorized patch panel has one or more multi-circuit connectors mounted at the rear of the panel. The modular jacks are each connected to the appropriate pins on the connector. Six 4 pair, 8-pin modular jacks can be connected to a single 50-pin "telco" connector (also called an Amphenol connector), if all pins of the jack are connected. This means that you would have to use four 50-pin connectors for a 24-position patch panel. Some patch panels connect only two pairs to each jack position and support 12 jacks per 50-pin connector. These panels are wired to support only one style of network connection, such as 10BaseT (which uses Pins 1, 2, 3, and 6) or Token-Ring (which uses Pins 3, 4, 5, and 6). Obviously, the styles may not mix, so you may have to replace the panel if you change your network hardware.

The mapping of connections from the connector to the jack is critical, as each pair on the jack generally has a preassigned circuit function, such as Transmit $+/-$. If used to provide connectivity to a multiport hub card, the connectorized patch panel takes the place of the modular equipment jacks. In a conventional modular jack hub, the hub connector pins are each assigned a specific network function, and any 4 pair equipment cord wired straight-through will automatically put the proper signals on the pins of the horizontal patch panel (and thus the cable). If a connectorized patch panel is used, the signals are mapped from a multicircuit connector to each modular jack on the panel. Instead of an equipment cord from the hub to the horizontal patch, a patch cord is connected from each hub-port jack on the connectorized patch to the corresponding station jack on the horizontal patch.

If the mapping from the 50-pin connector to the jack is wrong, the signals will be reversed or totally miswired and the connection will not work. The standard mapping for connections between the 50-pin connector and each jack is dependent upon pair numbering. It is customary to connect the pair color codes in the order given in Chapter 3 for the wire pairs. In other words, pair 1 of a 25 pair (50-wire) cable connects to Pins 1 and 26 on the connector, pair 2 connects to pins 2 and 27, etc. However, the connections do not go to a 25 pair cable but lead instead to individual jacks, either four or eight wires at a time. So a 4 pair jack panel connection would have pairs 1 through 4 connected to the first jack, pairs 5 through 8 to the second jack, and so on, in turn, up to the sixth jack (all that can be served by one 50-pin connector).

Then at each modular jack, the pairs are connected in a pattern consistent with either T568A or T568B. This distinction is important, as the two patterns reverse the connections for pairs 2 and 3. Remember that T568B is equivalent to AT&T 258A and is probably what you will get if you don't specify. You should always specify the patch jacks and connectors to be mapped to T568B or T568A, whichever matches your hub cord. You are still free to use a different pattern for the horizontal cable and patch, since it is separate from the hub-to-patch connection. Panels that use all eight pins of each jack are clearly identified as to pattern mapping, but panels that use only four pins use a skeleton of one of the TIA patterns. As we said, hub connections often use only two pairs per port (four wires) to maximize the number of ports per connector.

Here is an example of the problem of connection mapping. Let's say you intend to install a 10BaseT network using hubs with multicircuit connectors instead of single port jacks. Such hubs typically use a card rack with plug-in cards. You will run cables from your hub cards to a patch panel to fan out the individual ports to modular jacks. Naturally, you will want the 8-pin modular jacks on the patch to have the same pinouts as a standard 8-pin modular jack on a hub with individual connectors. A standard 10BaseT modular hub jack defines pin 1 as transmit $+$, pin 2 as transmit $-$, pin 3 as receive $+$, and pin 4 as receive $-$. A multicircuit hub connector could assign eight pins to each port, but only four pins are needed for 10BaseT, so assigning four pins per port achieves twice the density on the connector. For our example, let's say that your hub card connector assigns four pins per port, representing two pairs of the multiconductor cable. This gives you a total of twelve 10BaseT ports per 50-pin connector. The cable connects between the hub card and the patch.

Now, let's look at the way the ports are mapped from the 50-pin connector to each modular jack on the patch panel. Standard 10BaseT uses four pins of the 8-pin jack. These four pins correspond to two of the regular 4 pair groupings of the jack, pairs 2 and 3. Remember that T568A and B simply reverse the order of these two pairs. The "first" pair for each of the ports on the 50-pin connector usually corresponds to "transmit $+/-$". Now, here is the mapping part: the transmit $+/-$ pair must connect to pins 1 and 2 on the 8-pin jack, while the receive $+/-$ goes to Pins 3 and 6, for proper 10BaseT connectivity. If the patch uses the T568B pattern to connect the patch jack, the transmit and receive pairs will appear on the proper pins. But if the patch is wired for the T568A pattern, the two pairs will be in reverse order and the connection will not work (unless the pair order on the 50-pin connector has also been reversed—a rare circumstance because the TIA recognizes only the standard color code pattern for that connector, not a pair order).

This is one place where the TIA's preferred T568A pattern will get you into trouble. Note, however, that all of your modular patch cords, horizontal cable patch panels, and workstation outlet jacks can still use the A pattern. The rule is that both ends of a connection must be the same, whether that connection is a horizontal cable, a patch cord, or (yes) a patch panel fanout of a multicircuit hub port. Always ensure that you match the LAN connections on the hub card's connector with the connector-to-jack mapping on the patch panel so that the correct LAN signals will be on the pins of the patch panel jacks.

Connectorized patch panels may be rated by category as with any connecting hardware and should be properly marked. The actual internal wiring between the connector and the jacks must be appropriate for the category of operation, such as Category 5. The manufacturer will normally provide testing and verification by an independent laboratory, such as UL, to attest to their claimed category of operation. As with regular patch panels, connectorized panels that are not marked or otherwise certified, should not be used in any application that requires Category 4 or 5 performance.

Location and marking

Patch panels are usually designed to be mounted in 19 inch relay racks. They also mount on simple, freestanding mounting rails, in equipment cabinets, and on wallmounts. You should carefully plan the location of your patch panels, as you should all your connectivity hardware. If you choose to use patch panels, rather than direct crossconnection, you will have also to plan for routing and dressing of the patch cords from origin to destination jacks. If you route the cords using wire management brackets and accessories, you will avoid the "cable curtain" that sometimes infects patch panel installations.

As we said before, patch panels may be used for both the horizontal cable and the equipment (hub) port connections. If you use patches for both, you will have to provide a wiring closet layout that brings the two types of panels conveniently together for reasonable patch cord lengths. If your installation involves the frequent reconfiguration of networks and workstations, you should definitely plan to have the two panels as close together as possible. In large installations, each type of patch location may actually be composed of several panels. You may want to place them on multibay relay racks (side-by-side) so that patch cords can go directly across between racks to make connections, rather than climbing the rails in a route across the top of the racks. Wire management accessories for adjacent racks are available to make this job easier.

If you terminate the horizontal cables in punchdown blocks and then crossconnect to patch panels, you should plan the layout in such a way that you minimize both the length of crossconnect wire and the length of equipment cords (or patch cords) to the network hubs. You can easily mount punchdown blocks and patch panels on wiring closet backboards (wall mount), but the hubs will probably mount in an equipment rack, unless the installation is very small. This may increase the length of the cords from the horizontal patches to the equipment connections. Proper planning will minimize this distance.

You should carefully plan your wiring closet installation, including the location of the patch panels and hubs, to keep the total crossconnect and cord lengths within the TIA 568-A guidelines. You will recall that the maximum length of horizontal cabling is 90 m (295 ft). An additional length of up to 10 m (33 ft) for each link is allowed for horizontal crossconnect, jumpers, patch cords, patch panels, and user/equipment cords. That makes the recommended total length for all of the wiring in a horizontal link (Channel) 100 m (328 ft). This is the distance from the workstation, all the way to the final hub or equipment connection in the wiring closet, including the user cords on each end. (The actual connectors at each end are excluded in the Channel measurements.)

The standard also recommends maximum lengths cable in the work area and wiring closet which make up the 10 m (33 ft) allowance. For example, the total length of the horizontal crossconnect facilities should be no more than 6 m (20 ft). This distance includes the crossconnects, the patch cords or crossconnect jumpers, and the equipment cords in the wiring closet. The standard also recommends that the user cords in the work area be no more than 3 m (10 ft). Combining these two lengths should yield 10 m (33 ft), although the rounding-off of conversions by the standard makes the math look rather strange.

Work area cords of only 3 m (10 ft) is rather short for many office environments. As these cord and jumper lengths are only recommendations, you could adjust them to fit your needs, as long as you keep the total length under 10 m (33 ft). However, the only way to guarantee that you can make this tradeoff is to carefully plan the wiring closet installation so that you minimize the patch cord and equipment cord lengths. That will give you more length for user cords in the work area.

What difference does it make if you exceed these lengths a little bit? If the total Channel length is less than 100 m (328 ft), you will probably have no problems; however, if you run a longer link, it might not work properly. Network equipment manufacturers design their network adapters and hubs to operate over the standard link distances and characteristics. If you exceed those distances, or in other ways provide a substandard link, the networking components may not operate on that link. Also, keep in mind that some of the performance parameters of patch cords and crossconnect wire are worse than regular horizontal cable, so you may not be able to just borrow some of the allowed horizontal cable length and add it to the patch cord allowance.

In addition, a link that is too long will fail a cable scan. Just how long is too long? Cable length measurements are a part of the tests done by a cable scanner. These test procedures are supplied in a supplement to TIA 568-A called TSB-67. TSB-67 has length test limits for both the Basic Link and the Channel. If you exceed the test limits, the tester is required to fail the link. The test requirements for cable links are fairly detailed and are covered in Chapter 12.

If you test the part of the horizontal link called the Basic Link, it excludes the user cords, jumpers, and patch cords. When these cords are added in, the total link is defined as a Channel. You might later add cords that make the Channel exceed the allowable length, even though the Basic Link test was passed. If you exceed these lengths, your network equipment may not operate properly and it will be nobody's fault but your own.

Proper marking of the patch positions is also an important part of an installation. Generally, each wiring closet, cable termination, rack, patch, and hub is required to have a unique identification, according to EIA/TIA 606. In the case of patch panels, you will have to provide your own circuit or jack position identification separate from the panel's numbering, unless you have a very small installation and can use the jack numbers as your station numbers. The detailed requirements for circuit identification have occurred fairly recently, and it is also difficult to provide labeling for closely-spaced jacks, so many patch panel manufacturers just provide a stick-on label holder that can be positioned above or below a row of jacks.

A typical patch position might need to be marked with a wiring closet number, patch number, cable number, and jack position number such as "W4-P10-C321-J15." However, the space allowed for each jack might require this to be shortened to "C321-J15," with the patch number marked on the panel itself and the closet number obvious from where the panel was located. The rules for marking identification numbers are part of an emerging standard and will get more detailed as time goes on. Panel manufacturers will begin to incorporate more of the marking guidelines in their product's features. For the time being, we may have to make do with adhesive label strips and other marking methods.

Patch Cords

A patch cord consists of a length of flexible cable, terminated at either end by an 8-pin modular plug. The patch cord, in reality, is identical to the user cables in the work area and the wiring closet. As a matter of fact, all of these cords go by a number of names that reflect their purpose, rather than their construction. For example, a cord from a patch to another patch is called a *patch cord*, but an identical cord from a patch to a hub is technically an *equipment cord*. The cord from the workstation to the outlet connector is also an equipment cord. Equipment cords are sometimes called *user cords*, because they are presumably connected by the user, not the cable installer, to the user's equipment. Oh, yes, all of these cords are sometimes called *cables*, instead.

The one distinguishing feature of the patch cord is that it is usually shorter than the user cords. We will describe the wire and connectors for these types of cables in the next chapter. In this section, we will concentrate on the characteristics that make a good patch cord, with the understanding that these same characteristics apply to all user cables.

Now that we have all that terminology out of the way, what makes a good patch cord? As we mentioned, a patch cord should be flexible—which means it should use stranded wire with a flexible plastic jacket. Patch cords usually consist of four pairs of 24 AWG, 100 ohm, stranded copper, thermoplastic insulated wires with an overall thermoplastic jacket. As with horizontal cable, 22 gauge wire is allowed but rarely used. The plastic insulation is often PVC or a compound with similar characteristics. The cords are used in a work area, not a plenum space, and do not require the stiffer plastic insulations that are plenum-rated.

Patch cord cable may have any wire colors, but the colors of the standard 4 pair color code are normally used. TIA 568-A also offers an alternate color code, consisting of eight unique solid colors. Actually, the color code is only important if you make up your own patch cords from bulk cable. The colors will help you get the wires in the correct pin positions before you crimp on the modular plug.

Patch cords have their own set of performance characteristics that are slightly different from those for the horizontal cabling. Most of the transmission requirements are the same, but an additional 20% increase in attenuation is allowed for the stranded construction and "design differences." This is one of the reasons that the lengths of the patch and user cords are limited by the standards. The attenuation limits vary for the three categories of performance and are prorated from the values per 100 m (328 ft). The allowable attenuation values for patch cords are shown in Table 9.1.

If you purchase preassembled patch cords, you should be sure that the cords are certified by the manufacturer to comply with TIA 568-A for the appropriate category of performance. Certification testing by an outside agency, such as UL, is a good indication of quality and assurance. The testing of patch cords presents a somewhat of a dilemma, both for the manufacturer and for you. The standard contains detailed performance requirements for the cable and the

TABLE 9.1 Patch Cord Maximum Attenuation Values

Frequency	Cat. 3	Cat. 4	Cat. 5
0.064	1.1	1.0	1.0
0.256	1.6	1.3	1.3
0.512	2.2	1.8	1.8
0.772	2.7	2.3	2.2
1	3.1	2.6	2.4
4	6.7	5.2	4.9
8	10.2	7.4	6.9
10	11.7	8.3	7.8
16	15.7	10.7	9.9
20	n/a	12.0	11.1
25	n/a	n/a	12.5
31.25	n/a	n/a	14.1
62.50	n/a	n/a	20.4
100	n/a	n/a	26.4

NOTE: Attenuation values above are given in dB per 100 m (328 ft) at 20 degrees Celsius. The measured attenuation of a particular cable will depend upon its actual length, but the value (using the worst case pair) should be less than the prorated value at 100 m. The values in the chart are calculated from a formula at specific frequencies. The allowable attenuation of a stranded wire cable is 1.2 times the value of the formula in Paragraph 10.2.4.6 in TIA-568A, prorated to the actual length of the cable under test.

connectors that are used to make up patch cords, but there are no specifications for assembled cords. In addition, certain of the tests, such as NEXT, are known to be unreliable for links less than 15 m, because of a phenomenon called resonance. Many testers will not test a cable that is less than 6 m, except for wire map. Because of these factors, the only assurance of patch cord performance you will have (until the cord becomes part of a larger link) is the use of quality components and cable, and good workmanship. Workmanship is important because the cable pairs must be untwisted before the modular plug can be attached. If the amount of untwist is not minimized, the plug will contribute even more degradation to the link than is inherent in its design. By the way, the unavoidable amount of untwist in the modular connector design is one of the reasons that these connectors are not looked upon favorably for high-frequency performance. Annex B of TIA 568-A shows a detailed procedure for assembling modular cable ends. Although this procedure is for the preparation of connectors that are to be used in the testing of connecting hardware, it represents an ideal method to maximize the performance of patch cord plugs.

Fiber Optic Patches and Cords

Fiber optic links may also use patch panels and patch cords in the wiring closet. The very nature of fiber optic interconnection makes almost every fiber connection an optical patch, since the light path must be preserved. Optical cables from workstations or from other wiring closets typically terminate in termination boxes that allow excess fiber to be protected from accidental damage. The termination boxes may serve as the patch point for a small number of connections, but larger installations will have a separate patching location that serves all of the incoming and outgoing fiber cables.

Optical patch panels that are designed for TIA 568-A compatibility use the very same passive duplex SC adapter that is used at the work area outlet. The fibers of optical cables are terminated directly into SC connectors (or other types), rather than passing through some intermediate device, such as the connecting block used with copper wiring. The fiber terminations and connector types are described in detail in Chapters 7 and 8. We will concentrate on the physical arrangement of the patch panels and the characteristics of fiber patch cords in this section.

A fiber optic patch panel consists of an array of duplex SC adapters or hybrid adapters. If the entire installation, including the fiber optic hubs, repeaters, or network adapters, uses the recommended SC fiber connector, then the array will be made up of SC adapters. However, it may be necessary to convert fiber connections between other connector types. There are hundreds of types of LAN equipment with fiber optic interfaces that use one of the other connector types, such as the popular ST or SMA connectors. For several years, there will be a need to provide a conversion between the various connectors. The TIA 568-A standard contains suggestions for migrating to the SC connector system in Annex F. The reason the standard has chosen to recommend the SC connector (one that is relatively new) is that its square, tapered design allows quick connection and disconnection and is easily combined in polarized pairs.

To convert from another fiber connector type to the SC connector, you need to use a hybrid adapter or a conversion cable. A hybrid adapter is a passive coupler that joins two different connector types, while a conversion cable simply has one connector type on one end and the other type on the opposite end. For example, to convert an ST equipment interface to SC, you need a fiber cable with one of the connectors on each end. The duplex (two-fiber) cord described later is convenient to use, because you will need two fibers for each link connection. The other types of fiber connectors usually have no built in polarization, so you will have to pay attention to which connector goes to transmit and which to receive. The standard, at present, gives no guidance as to which A or B position of each orientation (AB or BA) should contain the transmit signal and which the receive. The designation of one fiber as transmit and one as receive is a little difficult anyway, because the transmit from the one end of a cable is the receive from the other. You will have to create your own standard practices for which connector position at the patch panel is transmit or receive.

Fiber optic patch cords are usually made from flexible duplex fiber cables, often referred to as "zip cords" because of their resemblance to the household electrical cords by the same name. Fiber zip cords differ in that the conductors are replaced by fibers, insulation does not have to meet any electrical specifications, and color is usually bright. Because the fibers are always patched in pairs, the dual-fiber construction is handy in that it easily splits out for two individual connectors.

A duplex clip can be used to attach the two SC connectors together. Fiber connector pairs that are thus clipped together have an orientation just as the adapters do. This is so that the A and B connections will be placed into the proper position in the mating adapter. There are two orientations for the connectors of a fiber patch cord, the AB and BA orientation, which were described in Chapter 8. It does not really matter which orientation you use for each end of the patch cord, as long as the two ends of the cord have opposite orientations, and the fiber polarity is flipped between the two ends.

Adapters and connectors may be marked with a distinctive color, such as red or white, to assist in the installation and later connection of mating plugs. The TIA standard does not specify or define these "shadings" at the present time, but you can use them in your facility to help differentiate the A and B positions of the mating connections. Remember that neither the A nor B designation specifies the direction of the optical signal but merely act similar to pin numbers in a conventional connector. Furthermore, since every fiber cable and adapter coupling reverses the A/B order, you could not tell which fiber carried which signal, even if you knew how the signals started out unless you traced every fiber through all the polarity reversals. As a user, the only thing that you would know is that the reversals had been arranged in such a way that whatever signal started out on the A position at one end of the link will be on the B position at the other end. This is done to create the transmit/receive reversal that must be done for a fiber link to function.

Remember that some fiber optic links may use laser illumination that is potentially hazardous. Most LAN connections that you encounter will use optical power levels far below that which would actually cause a problem, but you should still use reasonable precautions when working around fiber. As we covered in Chapter 8, you should never look directly into an open fiber connector (or an unterminated fiber that might be part of an active circuit). The light wavelengths used for these signals are in the infrared part of the spectrum and are quite invisible to the eye. Some fiber transmitters that are intended for long distances may actually have levels of light that would be hazardous, so just don't get into a bad habit.

10

User Cords
and Connectors

This chapter will describe the various types of connectors that are used to terminate cable ends. We have previously covered station cable termination, so this section will concentrate on the modular plugs, coax cable-end connectors, and Data Connectors that are used on freestanding cables such as patch cords and user cords (sometimes called user cables). We will also look at the user cords that are used in the work areas and the equipment rooms for connection to network hardware.

This is a good time to point out the distinction between "cables" and "cords" as used here and in many of the standards. Obviously, both cables and cords used in telecommunications employ jacketed, multipaired cables, but it is customary to refer to the flexible, stranded-wire, plug-terminated cables that connect to workstations, hubs, and patches as "cords." This is the convention followed by most of the standards. The usage rule is probably more consistent when referring to patch cords, which are inevitably referred to as "cords." The terms "user cord" and "user cable" are used interchangeably, as are the terms "equipment cord" and "equipment cable." For consistency in this book, we have tried to use the word "cable" to refer to any solid-conductor cables that are terminated in a jack or punchdown, and use the word "cord" to refer to any stranded-conductor cables that are terminated in modular plugs. With regard to coax cables, they are normally referred to as "cables," regardless of their location. By the same token (pun intended), IBM Data Connector cables are normally referred to as "cables." Both of these latter cable types are identical in composition and termination whether they are used in the wall or by the equipment user.

Cable-End Connectors

We will cover three general types of cable connectors in this section: modular connectors, coax connectors, and IBM Data Connectors. Fiber optic connectors have been described in Chapter 8 and will not be covered here. Of the three other types of connectors, the modular connector is obviously the most favored for modern LAN wiring, because of the increasing use of unshielded twisted pair cable. Coax has long been used for traditional Ethernet and Arcnet LANs, but is gradually being phased out in most locations. The Data Connector is an important connector in shielded twisted pair wiring systems and it is now included in the TIA 568-A standard.

RJ-45: What's in a Name?

The 8-pin modular connector is often improperly referred to as an "RJ-45." The terminology is in such wide use, that even manufacturers who know better frequently use the RJ-45 name in their marketing materials. Some will call it RJ-45 "style" or "type." In truth, the RJ-45 designation is for a particular interface most often used for programmable analog modem connections to leased telephone lines. This true RJ-45 just happens to use the 8-pin (8-position, 8-contact) modular jack and plug that are found in many other LAN and telephone connections that have nothing whatever to do with the original intent of the RJ-45 connection. Many other common interfaces use the same type of modular connector, including RJ-48, 10/100BaseT, Token-Ring/UTP, UNI-PMD, T1, ISDN, and on and on.

This common usage error would matter very little if you were never going to use high-speed data networks. However, the TIA 568-A version of the 8-pin modular connector interface has very specific performance requirements that may not be met by the generic nonrated connector that simply goes by the RJ-45 name. It would have been convenient if the TIA had given this connector a new name commensurate with its specifications. Perhaps it could have been called the "T568-8/Cat5" connector, or how about just "T-45" to borrow from both the past and the present.

When will people stop calling it an "RJ-45"? Don't hold your breath! Don't grate your teeth either! Just get used to thinking that a very special type of connector is meant when a "RJ-45" is used in your LAN wiring system. And, oh yes, never ever use the "RJ" word on the Net!

Modular connectors

The modular connector used in LAN systems is the same connector that is used in modular telephone wiring systems. This connector is available in several possible sizes and pin configurations, ranging from four to eight positions, with from two to eight contact pins. The popular styles are often referred to by a USOC (Universal Service Order Code) "RJ" number, even though the connector may not actually be used in the designated application called for by the USOC code. For example, the common 6-pin modular telephone plug is often called an RJ-11, and the 8-pin modular plug is often called an RJ-45. The 8-pin modular plug is used in the TIA 568-A wiring standard for both telephone and data. The 8-pin modular plug is also the connector that is used for 10BaseT, 100BaseT, 100VG-AnyLAN, Token-Ring/UTP, and many other LAN applications.

This 8-pin modular plug is probably the most subject to name abuse, because it resembles the specialized RJ-45 connector. However the RJ-45 wiring pattern (which includes an interface programming resistor) is so radically different from that of T568A and B that it really should not be called by that name at all. In this book, we have tried to always use the "8-pin modular," or simply "modular," description, as it leads to less confusion. In truth, the plugs are identical until terminated. Then, they should probably be called by a name that reflects their wiring pattern. For example, you could certainly connect only four pins of an 8-pin modular plug in a 10BaseT pattern. It would not truly be either T568A or B, because both require eight connected pins. It would not be an RJ-45, because the wiring pattern would be wrong and the programming resistor would be missing. Why not call it a 10BaseT connector, as that would be its purpose and its unique wiring pattern? If all eight wires are connected in an approved pattern, you could call it a "TIA" connector, or even "T568A or B." Well, until the habit is changed, we will probably still find people that call the connectors "RJ-45's," however wrong the term might be.

The 8-pin modular plug that is used in standard LAN wiring is specified by IEC 603-7, "Detailed specification for connectors, 8-way, including fixed and free connectors with common mating features." This is the plug that is specified in TIA 568-A and related documents, and in ISO/IEC IS-11801.

Modular connectors are primarily intended for terminating cables with stranded conductors. In fact, the original connector was designed to terminate a flat cable containing from two to eight stranded conductors. It was also designed primarily for the audio frequencies of telephone lines, although the connector is officially rated for use up to 3 MHz. Unfortunately, the industry not only uses the connector at frequencies far above that, but also needs to place twisted pair conductors, encased in a round cable jacket, into the modular plug. To allow the use of the modular connector at LAN frequencies from 10 to 100 MHz, the TIA has simply specified performance criteria (primarily attenuation and NEXT) that the connector must meet. As long as these criteria are met, the connector can be used in applications up to Category 5.

Modular connectors are available for solid wire, although some authorities discourage using solid wire for cords, even with the special connectors. The difference between the two connector designs is shown in Fig. 10.1. As you can see, the modular pin is a flat contact with a pointed end that pierces through the insulation of the wire and makes electrical contact with the stranded wires. The contact may have one or more points. If this contact is used on a solid wire, the conductor may slip off to the side of the points and make intermittent contact or no contact at all. For this reason, the solid wire contact has three pointed fingers at the bottom that are offset slightly, so that the wire is centered, the insulation displaced, and the wire held in constant contact with the opposing fingers. (You can also see why it is bad practice to use the wrong type of connector for the type of wire you are using.)

Shielded modular plugs have been developed that use a metal sleeve around the plug to provide shielding of the connection. These plugs require jacks that

Figure 10.1 The difference between the two modular connector contact designs for stranded and solid wire.

are compatible with the shielded plugs for proper functioning of the shield. Shielded cables can also have their shield drain connected to one of unused pins on the regular 8-pin plug, but the standard connection pattern of four balanced pairs is lost when this configuration is used. The only shielded connector that is recommended by the TIA is the Data Connector described later in this chapter.

The standard pin numbering and wiring patterns for the 8-pin modular connector were shown in Fig. 7.10. The T568A and T568B patterns featured in TIA 568-A are shown at the top (also in Fig. 10.2), along with other patterns for specific applications. The 10BaseT pattern is also applicable to 100BaseTX (but not T4 or VG, which require all four pairs). Notice that the T568A and B patterns actually support several other applications, including virtually all of the LAN connections for Ethernet, Arcnet, Token-Ring, and the PMD patterns of ATM. The key to whether an application can be supported is the pairing of the wires that connect to the plug. For example, the TIA patterns both put pairs on pin combinations 1-2, 3-6, 4-5, and 7-8, although in differing pair order. Some patterns, however, pair the wires 1-8, 2-7, 3-6, and 4-5, which places the outer four pins on different pairings. If your connection required pairings on a different pattern than was wired in the cable, the result would be a split-pair condition, which causes severe performance impairment above 1 MHz or so. That is why choosing the correct pattern for your equipment is so important. Fortunately, virtually all LAN equipment can be supported on either of the TIA patterns.

Figure 10.3 shows the recommended dimensions for an optimum modular plug termination. The dimensions are taken from Annex B of TIA 568-A. Notice that the crossover of conductor 6 occurs inside the plug, forward of the strain-relief tab. This is a little different from the standard practice, which places the crossover just outside the tab. Also notice the length of untwisted wire that exists inside the plug body. It is this untwist that hampers the performance of the plug in very high frequency environments.

Figure 10.2 The standard pin numbering and wiring patterns for the 8-pin and 6-pin modular connectors.

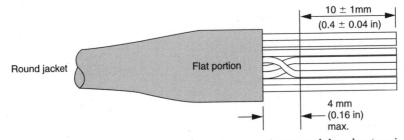

Figure 10.3 The recommended dimensions for an optimum modular plug termination. The dimensions are taken from Annex B of TIA 568-A.

Figure 10.4 Untwist the pairs back to the jacket, placing the wires of each pair in the proper order, according to the color code. Notice that conductor 4 crosses over conductors 5 and 6 inside the strain relief and terminates at position 6.

Figure 10.5 Typical crimp tool for module connectors.

The step-by-step procedure for terminating a cable in a modular plug is as follows. Refer to Fig. 10.3 and Fig. 10.4. Begin by removing the cable jacket a minimum of 20 mm (0.8 in) from the end of the conductors. Place the pairs in the order of the pins they will be terminated on (1-2, 3-6, 4-5, and 7-8). The color of the first two pairs depends upon whether you use T568A or T568B. Flatten the cable jacket at the end, to place the pairs side by side. Untwist the pairs back to the jacket, placing the wires of each pair in the proper order, according to the color code shown in Fig. 10.4. Arrange the wires so that they are parallel and lay flat. Cross the wire for pin 6 over the pin 4 and 5 wires such that the crossover is no more than 4 mm (0.16 in) from the edge of the jacket. Trim the conductors to approximately 14 mm (0.55 in) from the edge of the jacket. Place the plug over the conductors to that they extend all the way to the bottom of the termination slot and the jacket extends at least 6 mm (0.24 in) inside the plug body. Crimp the plug using a modular crimp tool. After both ends of the cable have been terminated, test the connections for continuity and proper conductor placement.

A crimp tool with a machined crimping die should be used for the best connections. A typical crimp tool is shown in Fig. 10.5. A proper crimp tool will cost as much as $100 to $200, but will be well worth the expense. The tool should have a ratchet action that will not allow the jaws to open until the plug has been fully crimped. Inexpensive tool designs that were adequate for the smaller 6-pin modular telephone connectors will typically fail to properly crimp the inside pins (pins 3, 4, 5, and 6) when the design is extended to 8-pin widths. This may or may not cause immediate failure of the connection, but you will have problems over time, even if the connection initially tests good. You can inspect a plug to see if it is properly crimped by looking at the plug end. A properly crimped plug will have all eight contacts fully engaged and even on the tops, which will be very slightly below the plastic channels between the contact pins. An improperly crimped plug will show a distinct rounding upward of the center contacts. Side-action crimp tools may favor the contacts on one side of the plug. The plug should also be inspected from the front and top to ensure that the contacts have not drifted to the side, as can happen with inexpensive modular plugs. As simple as these plugs are, there are still significant differences in quality.

It would not be proper in a discussion of LAN cable terminations to ignore the problems that result from the choice of the classic 8-pin modular plug and its mating jack as the standard UTP connector. These components are responsible for the majority of limitations with this generation of high-speed networking. The connector, as we said before, was designed for frequencies well below 3 MHz, and it has very poor characteristics at frequencies approaching 100 MHz. The NEXT performance of the plug and jack is so bad at 100 MHz that you will see few manufacturers quote performance margins that exceed the TIA performance standards by more than 1dB, if they quote a margin at all. Test equipment manufacturers even create specialized low-NEXT adapters or digital processing techniques to compensate for the performance deficiencies of the modular connector in their own test equipment.

Nevertheless, with quality materials, proper assembly procedures, careful workmanship, and proper testing, the modular connector can support the highest performance metallic cable networks in wide use today.

Coax connectors

Coax connectors have been used for LAN connections for a very long time. The two most common LAN coax connectors are the BNC and the Type N connectors. The Type N was used on the original "thick Ethernet" or thicknet cable that first introduced Ethernet to the world. The connector is designed for the large cable diameter of coax cables such as that used for thicknet. The Type N is shown in Fig. 10.6. Type N connectors are expensive and bulky, but they are commonly available and are very reliable. The connectors are designed for use at many times the frequencies of Ethernet LANs, so their use in networks is no problem. The use of thicknet cable and the corresponding Type N connector has decreased greatly with the advent of thinnet cabling and its BNC connectors.

BNC coax connectors are smaller, less expensive connectors compared to the Type N. BNC connectors are used with "thin Ethernet" or thinnet cable that was introduced during the 1980s as a cheaper alternative to thicknet. The BNC connector is also used for Arcnet coax connections. The typical measurements to prepare a coax cable for BNC termination are shown in Fig. 10.7. BNC connectors are available in two-and three-piece constructions that are designed for termination using screw-on, crimp-on, or solder assembly.

Figure 10.6 The Type N coaxial connector.

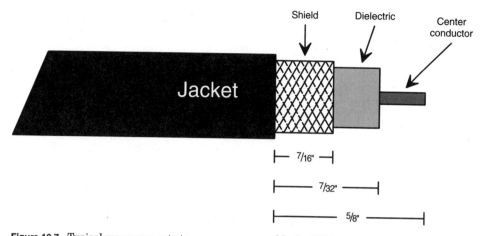

Figure 10.7 Typical measurements to prepare a coax cable for BNC termination.

The crimp-on style is the most popular for field termination of LAN cables. The assembly process is as follows. The crimp-on sleeve (of a three-piece connector) is placed over the end of the coax. The coax cable is stripped in the manner shown in the figure, exposing the center conductor and the braided shield. The center pin is crimped onto the center conductor. Next, the connector body is slid over the coax so that the back barrel of the connector slides under the braid and the center pin is properly positioned in the insulator in the center of the connector. The sleeve is slid down over the exposed braid until it contacts the connector body. Finally, the sleeve is crimped onto the connector using an appropriate crimp tool. The assembly is tested for electrical connectivity and mechanical soundness.

The proper tools are essential for anyone doing coax connector termination. A rotary coax stripper makes preparation of the cable ends a snap and eliminates the need for measurement. The tool has two blades that cut through the insulation to the precise depth to expose the center conductor and the braid. Crimping the center conductor and the outer sleeve also require the proper crimp tool. As with crimp tools for modular connectors, the best tools are machined-die tools with a ratchet-type action that ensures the tool is completely closed, and the sleeve or pin fully crimped, before the handles release.

A properly crimped connector can easily withstand a moderate amount of stress without pulling loose. If you use plenum-rated cable, be aware that the dimensions of this type of cable are much smaller than standard non-plenum coax. In addition, plenum coax dimensions may vary slightly with different manufacturers as the dielectric insulation and jacket composition may differ. It is essential that you use coax connectors that are designed for the style of coax you are using. Plenum-style connectors have a smaller barrel that matches the smaller diameter of the dielectric layer of plenum coax.

IBM-style Data Connectors

The hermaphroditic IBM Data Connector is the recommended connector for 150 ohm shielded twisted pair (STP) cabling systems. It has long been standardized in IBM, ANSI, and EIA documents and is now a part of TIA 568-A. The connector design, shown in Fig. 10.8, is defined in *IEC 807-8, Rectangular Connectors for Frequencies Below 3 MHz,* with additional transmission requirements in ANSI/IEEE 802.5 (ISO 8802.5) and TIA 568-A.

Although the title of the IEC document implies a 3 MHz frequency limit, the connector is actually designed for much higher frequency operation. The TIA 568-A characterizes the connector parameters to 300 MHz, three times that of the humble 8-pin modular connector. The connector incorporates an electromagnetic shield to complement the shielded cable structure. This connector could theoretically support data rates beyond the current 100 MHz offerings.

The Data Connector is configured for four signal wires and a shield. The connector design uses IDC contacts for the cable wires. The contact pins are color coded red, green, orange, and black to match the color coding of the STP cable. The wires terminate in the pins with a plastic clip and the cable body snaps together, making the installation almost toolless. A rotary cable stripper is handy to remove the outer jacket and trim the shield.

The step-by-step assembly instructions are as follows. Place the threaded connector sleeve over the cable. Then remove 1⅛ inch of the outer jacket. Trim the braided shield and the foil pair shields to within approximately ⅛ inch of the edge of the jacket. Place the cable into the connector body and bend the braid back against the connector shield. Route the wires to the proper color-coded termination posts, push the wires into the slotted jaws of the posts, and snap the retaining clip over the posts with the recommended tool or with a

Figure 10.8 The Data Connector design is defined in IEC 807-8 *Rectangular Connectors*.

pair of pliers. Place the side cover onto the connector body and snap it into place. Test the electrical integrity of the connector.

Two versions of the Data Connector are available that have slightly different mechanisms to lock the connectors together. They are referred to as style A (the original IBM version manufactured by AMP) and style B. You should decide which style works best for you and stick with it throughout your installation.

User Cords

User cords are the final connection step in a LAN wiring system. They are the cables that connect between the installed universal wiring system and the network equipment. They exist on both ends of a link, at the work area and in the wiring closet. In most modern LAN installations, the wiring closet will contain wiring hubs that consolidate all the station cables into a contiguous network. Wiring closets that contain network equipment in the form of hubs, repeaters, and servers are also "equipment rooms," which adds a few requirements over and above those for a simple telecommunications closet, but we will just call them wiring closets to keep things simple.

User cords include workstation cords and equipment cords. Technically, user cords do not include patch cords, which are part of the horizontal crossconnect. Sometimes all modular cords are called patch cords, regardless of their function, but we will make a distinction between true patch cords and user cords. Patch cords, which were covered in the last chapter, are essentially the same as user cords, except for length and color.

User cords are a required part of the Channel, the total end-to-end LAN connection, but they are not part of the Basic Link, which excludes user cords. These terms are defined in *TSB-67, Field Testing of Unshielded Twisted-Pair Cabling Systems,* and they are very useful in describing the component parts of a cable link. The theory is that a cable plant could be installed and tested before any of the user network hardware is installed. To do the testing, we need to have performance specifications for the portion of the horizontal cabling that excludes the user cables. That portion is called the Basic Link. However, Channel performance levels are also defined, so we could also test with the user cables in place. In fact, that type of test is the ultimate assurance that a link meets the requirements for a particular category of operation. Testing of installed cabling is covered in Chapter 12.

Construction

User cords are flexible, 4 pair cables, terminated at either end in an 8-pin (IEC 603-7) modular plug. The cords may be provided in various lengths, as needed for the application. The cords should meet the transmission and construction requirements for patch cords in TIA 568-A, as user cords are not explicitly defined elsewhere in that standard. The color code for the conductors and the wiring pattern connections for the modular plug are the standard configuration as shown in the first part of this chapter, as well as in Chapters 3 and 5. Either the T568A or B pattern may be used, as long as the same pattern is used at both ends of the cord. The standard also allows an alternate color pattern consisting of eight solid colors that can be used for user cords.

User cords are made from AWG 24, stranded, thermoplastic insulated conductors, arranged into four pairs, with an outer thermoplastic jacket. Conductors of 22 gauge wire may be used, if the cord meets the same performance requirements, but you must be certain that the connector plugs are designed for the larger wire size. The various jacketing materials are described in Chapter 3. Stranded wire is used because of its flexibility. From a technical standpoint, solid wire could be used, but it lacks flexibility and is more difficult to connect with modular plugs. In most locations, nonplenum insulation, such as PVC, is acceptable. PVC is also much more flexible than most plenum-rated plastics and it is more easily tinted.

The outer jacket color should be appropriate for the office environment. Some of the cable manufacturers offer Category 5 cable in a bright color, to distinguish it from lower category wire. There is no need to use this bright color in the work areas. The cable should be marked to indicate the category if there is any doubt. Compatibility with work area electrical cord colors is desirable for user cords. Many colors are available, but customary colors of beige (or ivory), gray, and black are commonly available for user cords.

Category

User cords have the same performance category designations as the other portions of the horizontal cabling system. However, since the 8-pin modular cords are very similar to nonrated telephone cords, you should be certain that the cords you use are certified to at least Category 3 levels. Cables are often called "data grade" by suppliers. This is an older classification system that simply indicates that the cables offer a little higher performance than "telephone grade" cables. Unfortunately, you need to know exactly how much better the performance will be, so insist on a proper certification of Category 3, 4, or 5, as needed.

Of course, if you are actually connecting telephone hardware, which is certainly permitted by the "universal telecommunications cabling," then telephone grade user cables are adequate. However, they do not actually meet the requirements of the TIA and other standards and may eventually get mixed in with the high-performance categorized cables.

User cords should be marked with the rated category of operation. The proper marking is "Category n" (n = 3, 4, or 5), "Cat n", or a letter "C" with the category number enclosed. Categorized cable is normally marked on the jacket, while the connectors may or may not have their own markings. Preassembled cable should be made from properly certified components, and should bear an additional marking indicating that the assembly meets workmanship standards or has been tested to the appropriate category.

Length and routing

According to the guidelines you must limit the length of your user cords in both the work area and in the wiring closet. The work area cords should be less than 3 m (10 ft), if possible. The cords in the wiring closet require a little simple math, as the length guidelines for the user cords are not separately specified. The total length of equipment cords, patch cords, and crossconnect jumpers in the wiring closet should be less than 6 m (20 ft). The length of the user/equipment cords will have to be taken from that total allowance. The standards allow 10 m (33 ft) for the all user and patch cords and jumpers. The two end-area lengths do not quite add up to this total because of rounding.

You might think that 3 m (10 ft) of user cable for the workstation might be a little short in some locations, and you would be right. This number is a recommended, not mandatory, requirement. The 10-m (33 ft) limit is mandatory, however, so if you need longer cables in the work area, you will have to subtract from the allowance in the wiring closet.

In reality, the total length of each horizontal link is the real operating limit of a LAN circuit. This limit is 100 m (328 ft) for the entire Channel, including all of the cables, cords, and jumpers that run all the way from the back of the workstation to the network hub. The 100 m (328 ft) includes 10 m (33 ft) for all the cords and jumpers, with the understanding that the stranded wires used in these cords have a 20% greater attenuation. Although the standard requires that all horizontal cables be solid wire, you could safely exceed the 10-m (33 ft) limit for user cords and jumpers by a little bit if you were well below the 90-m (295 ft) limit for the horizontal cable.

User cords in the work area should be routed with the same guidelines as for the horizontal cable runs. You should avoid sharp bends, kinks, and abrasions of the cable, and you should route it away from potential interference sources such as fluorescent lights and motors. Modular furniture usually requires that user cords be placed in raceways that are built in to the furniture. Remember that the guidelines suggest a separation of 2 to 5 inches from power conductors, such as the AC outlets (or mains) in the furniture channels. Furniture manufacturers are beginning to recognize the need for totally separate telecommunications raceways, so you may be able to specify furniture that meets this requirement. The placement of Category 5 cable is much more sensitive than lower categories. If you are operating a 10BaseT or Token-Ring 16 network, you really do not need Category 5 performance, and you may want to relax the rules somewhat. Routing of user cables in the wiring closet will not be covered here, as it is identical to patch cord routing, which was covered in Chapters 8 and 9.

Workmanship and quality

If you build your own user cords, or if you manufacture them for others, the quality of construction is a very important factor in the performance of the cord. You should first ensure that all connectors and cable used in the assembly of the cord are certified by the manufacturer for the level of performance you desire for the completed cord. Then you should use proper techniques to assemble the cord for maximum performance. It is known to be difficult to test short lengths of cable, particularly with regard to NEXT, so your best performance guarantee is good workmanship. Basic continuity and wire map, of course, can be completely tested without regard to length.

The section on patch cords in Chapter 9 and the earlier section on modular connectors in this chapter describes some techniques to ensure a high-quality termination that will perform up to expectations.

Wire Management

Locating Wiring Closets

The concept of a series of interconnected wiring closets is central to the structured cabling system design. The planning and placement of these wiring closets is a subject sufficiently important for the EIA/TIA to devote an entire standard to it, *EIA/TIA 569, Commercial Building Standard for Telecomunications Pathways and Spaces.*

In this chapter, we will discuss the placement of wiring closets in modern telecommunications installations. Naturally, this discussion will center around designing a wiring system for LAN use. This will impose some additional considerations on the location of our wiring closets, primarily because of the distance restrictions of LAN wiring.

The wiring closet is more broadly defined in EIA/TIA 569 as a hierarchical system of telecommunication spaces that are used for the concentration and connection of a building's telecommunications wiring. This standard covers wiring spaces for all telecommunications activities and actually includes telephone as well as computer wiring. Because of the much lower operating frequency, telephone wiring is not subject to the severe distance limitations of LAN wiring. For example, the standard allows a total distance of 800 m (2624 ft) for an unshielded twisted pair circuit run through a combination of closets—far beyond that allowed for a single LAN leg, which is limited to 100 m (328 ft) without passing through a repeater or bridge. So, we will concentrate on criteria appropriate for LANs, and assume that the telephone wiring will either easily meet its requirements or be separately handled.

In this hierarchical wiring closet system (so-called structured wiring), there are three types of wiring closets: the Telecommunications Closet (TC), the Intermediate Cross Connect (IC), and the Main Cross Connect (MC). Figure 11.1 shows this hierarchy. The structure does not assume that any of these closets necessarily contain equipment (such as hubs or telephone gear) but allows for any closet to be collocated with an Equipment Room (ER). In the traditional structured wiring installation for telephones, a cable from the local

ER	Equipment Room
HC	Horiz. Crossconnect
IC	Intermediate Crossconnect
MC	Main Crossconnect
TC	Telecommunications Closet
WA	Work Area (Outlets)
x	Crossconnect

Figure 11.1 The structured wiring hierarchy includes the Telecommunications Closet (TC), the Intermediate Crossconnect (IC), and the Main Crossconnect (MC).

exchange carrier (LEC) would enter the building and run to the MC, which would generally also house the user's telephone switching equipment (a PABX or EKTS). This would be a combination MC/ER closet.

Cross connections would be made between the LEC lines and the telephone switching equipment and additionally between the telephone switch and large multipair cables running to the ICs or TCs on each floor. A large building would have intermediate cross-connect (IC) spaces where the multipair cables would be terminated and split out to other feeder cables to TCs near various office locations. At the TC, the station cables from each workplace location would be cross-connected to the feeder cables. The line for each phone would actually be connected through the TC, to the IC, to the MC, and finally to the telephone switch. In a smaller building, the functions of the various closets may be combined. In a very small building, all connections might be made in one space, a combination TC/MC/ER.

LAN wiring is implemented on basically the same scheme but with several important differences. First of all, LAN systems are distributed systems. There is no large central device similar to the PABX from which all the connections emanate; instead, the connections are made to hubs located in the TC (thus an TC/ER). Then, as we will see, the hubs are interconnected to each other and to common resources such as servers. With regard to LANs, TCs are usually interconnected with no regard to hierarchical level, since the network components are distributed. Still, the wiring closet hierarchy is a useful model.

Horizontal and Backbone Wiring

All of the wiring from workstations to wiring closets and between closets is referred to as "horizontal" wiring and "backbone" wiring, respectively, in the terminology of structured wiring. Backbone wiring is sometimes called *vertical wiring*. If we envision workstations on each floor of a building, and wiring closets vertically stacked with one per floor, this terminology makes sense. However, multiple wiring closets are often located on the same floor, so the term "backbone" is clearer than the term "vertical"—the wiring from workstation to wiring closet is the horizontal wiring and the wiring between closets is the backbone wiring. If we also have additional wiring closets, such as the intermediate and main cross connect closets, the wiring between those closets is also called backbone wiring. This terminology holds whether the closets are on multiple floors or are on the same floor.

The EIA/TIA standards always use the term *backbone* to refer to wiring between closets. Unfortunately, backbone has a very different meaning in LAN systems as compared to telephone systems. A LAN backbone is a logical subdivision of a network that has nothing to do with where the wiring goes. In fact, wiring on the same LAN backbone segment can be in two or more wiring closets, or two LAN backbones can be in the same wiring closet. However, we will stick to the wording in the standards and refer to all wiring between closets as backbone wire (or cable), unless we are specifically referring to that portion of a LAN system that interconnects multiple LAN hubs or bridges.

EIA/TIA 569 recommends that the cross-connects in a wiring closet be grouped together and proscribes a color code to further show this grouping. Table 11.1 shows the recommended colors for various terminations and cross-connect locations. For the cross-connect locations, the standard generally uses traditional colors, such as orange for the telco demarcation punch blocks, and blue for the station cable punchdown locations. Actually, the punchblock itself is usually white, but its cover or mounting board may have the code color.

The standard also recommends that different jacket colors be used on the cables to distinguish between the usage types of the cables. This presents an

TABLE 11.1 Color Coding Areas in the Structural Wiring System

Orange	Telephone company demarcation	Blue	Horizontal
Green	Customer network connections	Purple	PBX, host, LAN, MUX
White	Backbone intra-building, 1st level	Red	KTS, EKTS
Brown	Backbone inter-building	Yellow	Misc.
Gray	Backbone intra-building, 2nd level		

NOTE: It is recommended by the standard that crossconnects be grouped by color for ease of identification. It is also recommended that different sheath colors (cable jackets) be used to aid in distinction between types of circuits.

interesting dilemma, as the colors in the standard bear little resemblance to the traditional colors used in commercial cabling. For example, station wire (to the workstation) has traditionally been run in beige or gray (the plenum-rated default), but the standard shows blue for this use. The color gray is used in the standard but only for second-level backbones, from the IC to the TC. Many LAN installers would agree that there is an advantage to having the LAN cables a distinctly different color from the telephone cables, but the standard assigns the same color to both types of station cables. (Advantage lost!) Of course, if the telephone cabling is simply run in a traditional color such as gray, despite the standard, then the LAN cables may be run in the recommended blue color to distinguish them from the telephone cables. However, in a design that includes both telephone and LAN cabling, you could not simply specify that the EIA/TIA standards be used in your statement of work, because the telephone cable color would not be in compliance.

Frankly, it would be easier to just use traditional colors for all the telephone wiring. Intercloset cables are easy to recognize because they have distinctly larger pair counts and are consequently physically much bigger than the station cables. You could adhere to the color code at all points of termination or interconnection for the telephone system. Since many cabling veterans strongly recommend keeping LAN wiring completely separate from telephone wiring, the use of contrasting blue LAN station wire is appealing. You could allow a deviation from the cable color recommendation for the telephone wiring but would maintain color coding for the LAN cable.

How many horizontal cables should you run between closets and to the work areas? The standard states that you should run two cables to each individual work area, but is this enough? The answer to this question depends upon many factors, but the two most important considerations are growth and utility. You must determine what amount of growth planning is reasonable, and what it will cost to implement. For example, if you have large individual offices that may eventually include two workers, overwiring by a factor of two would be reasonable. The type of work that is to be done is also important. For example, some workers may eventually need two computers or two phone lines. In such cases, you should provide at least one additional station cable outlet that can be used for expansion. Generally, it is far less expensive to do a certain amount of overwiring during initial construction than it is to add outlets later.

What methods should you use to terminate the cables in the wiring closet? The answer to this question depends upon whether you intend to install a universal wiring system, where any cable can serve your telephone or LAN systems, or install separate telephone and LAN wiring systems. Telephone wiring is quite forgiving in that you may run longer distances and make a larger number of cross-connects between the telephone instrument and the switching equipment. Modern LANs are much more restrictive in this regard. Early cabling standards allowed two cross-connects for a single LAN connection, but the latest Category 5 wiring standards (as shown in Fig. 11.2) allow only one

Interconnection

Crossconnection

Figure 11.2 The methods of interconnection and crossconnection are shown.

crossconnect and also severely restrict cable lengths, including that for the crossconnect wire. This is to allow for the strict attenuation, delay, and crosstalk limits of 100 MHz networks. Many experts even recommend that the workstation cables be terminated directly into a patch panel, with no intermediate cross-connects at all.

One of the reasons for this crossconnect limit is that all wiring terminations require that the pairs be slightly untwisted to be punched down at the point of termination. When the cable being used is very tightly twisted, as with Category 5 cable, this untwisted portion increases crosstalk, introduces noise, disturbs the line impedance, and increases RF emissions. Too much untwist can cause a link to fail at the higher LAN data rates, especially if the cable run is near the allowable length limit. All punchdown blocks introduce untwist into a link. The extent of the problem is less with the 110-style blocks than with the older 66-style blocks, but it exists nevertheless. The effect of untwist is more disruptive as the network speed increases. Terminating the station cables directly onto the patch panel minimizes the amount of untwisted wire and increases the performance of the link.

Distance Considerations

The primary guideline in locating wiring closets is to place them near the work areas they will serve. Advisory language in the standard says to place them in the center of the area to be served, but this central location is simply impossible in many existing buildings. The layout of the building may cause you to have to run your cables from a closet location on one side of the floor to each of the workstations. In such cases, it is critical that you predetermine the lengths of the wiring runs; if those runs will exceed the appropriate limits, additional wiring closets will be needed.

In new construction, the architects should be given firm instructions as to the placement of infrastructure facilities such as wiring closets and equipment rooms. If you make these needs known at the inception of building design, you will be more likely to have the telecommunication spaces that meet standards.

You should give consideration to placing the wiring closets vertically, one over the other, so that wiring distances between closets will be minimal. In large buildings, multiple "utilities shafts" not only simplify telecommunications wiring but also eliminate the single point of failure in single-shaft designs. The benefits of this approach go beyond the telecommunications issues. It is also a good idea to provide an orderly system of cable ducts or sleeves, both for inter-floor and firewall use. Proper ducting will allow workers to place UL-rated firestopping material in the ducts to enhance fire protection. Also, because many of the utility shafts are closed and reinforced concrete structures, this avoids the problem of random wall penetrations and "sneaker" cables that are forced alongside existing pipe and HVAC penetrations.

Maximum drop length

LAN wiring standards dictate strict distance limits for horizontal cable runs. These cable drop lengths include the entire cable from the point of termination in the wiring closet to the workstation outlet. Standards also cover the maximum distances between telecommunications closets. In this way, the requirements of TIA 568-A and 569 work together, in balance with the building's architecture, to achieve a compliant wiring system for your LAN that will meet your needed performance.

The maximum length for a horizontal cable is 90 m (295 ft). This 90 m length excludes the user cord that connects from the outlet to the user workstation (or telephone instrument) and the patch cord (or jumper) that connects from the patch panel (or crossconnect field) to a hub or other device. This is a maximum acceptable length and not simply an average length or a goal. You should try to keep your wiring well below this length in most cases. At least two telecommunications outlets must be provided for each individual work area.

Although some of the lower speed LANs, such as 10BaseT and even Token-Ring, may run to longer distances of as much as 150 m, exceeding the standard intentionally is a bad idea. You cannot expect such run lengths to support 100 Mbps data rates at all and any equipment problems will be blamed on the excessive drop lengths. As a matter of fact, cable testers are required to fail a Basic Link longer than 90 m, allowing for test cables and measurement accuracy. For a complete discussion of measuring cable lengths using cable testers, see Chapter 12.

The allowable backbone cable distances between wiring closets depend upon the type of cable you use as well as practical considerations. For example, the standard allows 2000 m (6560 ft) from TC to MC for fiber optic cable, but only 800 m (2624 ft) for unshielded twisted pair cable. The distances are even less between ICs and the other two types of closets. However, practice may dictate much smaller distances. If you wish to connect between two wiring closets on a 10BaseT or similar link, for example, you would have to limit the backbone run to the same 90 m you are limited to for any hub-to-workstation link. As mentioned previously, although some situations may operate with slightly longer cable lengths, they do not meet the requirements for inevitable higher speeds of 100 Mbps or more.

Estimating drop length

Cable drops from the proposed wiring closet to the workstations to be served must be within the maximum allowable length of 90 m (295 ft). You should estimate these distances before finalizing your cable system design. If the estimates exceed the maximum length, you may need to add one or more wiring closets. Although there are minimum size specifications and features for wiring closets (covered in a later section), you may be justified in putting in a substandard concentration point to get the drop lengths down.

In order to estimate the total length of a cable run, you must first know how the cable will be routed and hung. Conventional wiring is often routed through metal floor or roof supports to keep it well above the ceiling tile. Alternately, cable may be hung with a variety of cable hangers, clips, and tie wraps from supporting structures. In some installations, particularly old work, the cable is laid on top of the ceiling tile supporting grid. Large buildings frequently have the cable placed in cable trays suspended between the ceiling grid and the floor above. This method is becoming more popular for Category 5 installations where the slightest kink or sharp bend can impair the cable performance. Any combination of the methods may be used as well.

The routing methods are important because the more orderly routings inevitably require more wire. For example, routing through cable trays involves lots of right angles (orthogonal routing) instead of diagonal runs, demanding up to 40% more cable. On the other hand, laying the cable directly on top of the ceiling grid with direct routing uses the least cable (although it is the hardest to maintain and may suffer EMI problems more often). Cable trays are easiest to use during construction buildout, when the ceiling grid is not in place. They are more difficult to add once the offices are constructed or, worse, occupied. Trays carry significant weight from the many cables they contain and must be well secured in place. For that reason, cable hangers and clips are often used to minimize disruption to an existing workplace.

To estimate an orthogonal wire placement, where cables are to be bundled or laid in cable trays, you must calculate the length of run in the tray. This may involve a straight run or one with several corners. Add each leg of the run in the tray. Next, add the drop from the tray to the wall header above the workstation outlet, add the lengths of the vertical drops in the workstation wall and in the wiring closet, and add a factor for obstacles (2 to 5%, depending upon the building). Obstacles that might be encountered are HVAC units, duct work, firewalls, columns, and pipes. The vertical drops should allow for 18 inches (0.5 m) of excess wire at the wall outlet plus the distance through the wall to the height of the cable tray (not just the ceiling height). The vertical drop length in the wiring closet should include the drop from the tray to the final termination point of the cable, including all ducts, ports, channels, and trays through which the cable must be routed. Be sure to accurately estimate the twists and turns the cable may make to "dress" the wires on the mounting board or in the equipment racks. Allow another 6 to 12 inches here for termination. If your measurements are close to the limit, remember that cables are typically routed along the length of a patch panel to their termination points, which might add up to 19 inches of additional length.

Diagonal wire placement is really direct routing. This means that the cable is simply run above the ceiling grid and routed directly to the workstation drop location. Routing is done either directly over the grid or through supporting structures. Sometimes wire hangers or clips are used, as are tie wraps.

The use of cable hangers and clips may allow a combination between orthogonal wire placement and diagonal placement. Bundles of cable are typically routed through cable hangers orthogonally from the wiring closet to groups of work area locations, but individual work area drops cables may be diagonally routed from the nearest hanger.

Older buildings, houses, or buildings with fixed ceilings present special problems to cable routing. Often the cables may be run in attic or crawl spaces. Sometimes cable must be run around baseboards or even over door frames to reach the workstation location. The use of staples—a common mounting technique for telephone wires—is definitely out, even for Category 3 cable. Be careful when estimating cable lengths for these buildings. Simple detours, such as doorways, can eat up lots of cable. For example, a 10-foot by 20-foot room would require $10 + 20 + 17 = 47$ feet of cable simply to run to an opposite corner of the room, if one 7 by 3 ft $(7 + 7 + 3)$ doorway has to be traversed. If a drop must be run through several adjacent rooms to reach from the wiring closet to the work area, an enormous amount of cable may be required.

Accessibility, Appearance, and Access Control

The wiring centers of your network are key components of your LAN system, containing important networking hardware, the cable distribution points for your users, and important diagnostic and backup equipment. In addition, many important installation, configuration, and troubleshooting tasks take place there. For those reasons, you should devote the time and money necessary to maintain good wiring closets and equipment rooms.

The first key to a good wiring closet is accessibility. The wiring connections and network hardware should have their own room, out of the way and yet easy to get to. The area should be as close to the work areas as possible, since LAN workers will often have to travel between workstation locations and the wiring closet. That means that a location down the hallway near the elevators is not as useful as a small room or office near the work areas. The room should be used exclusively for LAN wiring terminations, hubs, routers, servers, and related equipment, but never for storage.

The second key is appearance. The wiring closet should be orderly, well laid out, and well lighted. Adequate power and ventilation should be available and designed for the specified purpose. Mounting boards on the walls should be full sheets of ¾ inch plywood and well secured to the wall's interior structure. At least two walls should be plywood-sheathed, and the boards should be painted a neutral color, such as gray. Areas for specific equipment should be well marked.

If you use equipment rails or racks to mount your hubs and patch panels, they should be placed a sufficient distance from the walls to work behind them. It is not at all a bad idea to place the hubs so they can be viewed from the doorway of the wiring closet. If you terminate the station cables directly onto a patch panel, you can use cable trays across the top of the mounting rails to route your cables. Wire management trays, mounting brackets, preassembled backboards, and distribution rings are all available to help make a neat and serviceable cable arrangement. Proper routing (or "dressing") of cables will make a neat, orderly appearance that will save maintenance time and trouble in the long run.

Category 5 wiring requires special precautions in installation and routing. The cable must be treated with great care during installation to preserve its performance characteristics. This special care includes running the wire with a minimum of sharp turns, no stretching, no staples, and no tight tie wraps. Many patch panels have additional strain relief points at each cable termination to eliminate sharp bends and to help secure the cables with a minimum of stress.

The third key to a successful wiring closet installation is security. Equipment located in the wiring closet needs to be secured. That is one reason why a separate room with a locked door is the best location for your wiring closet. If your wiring closet is located in a common area outside your offices, controlling access is essential. Wiring closets that are within your offices still may need to be secured, particularly if it is in a high-traffic location. Your company's mission may also dictate a need to control access to the closet, because important servers are often located there in addition to the networking hubs.

Size Requirements

Wiring closets need to be large enough to provide a proper area for wiring termination, patch panels, and hubs. In some cases, other networking equipment will be located in the same area. Rooms containing network hardware such as servers, bridges, and routers are properly called equipment rooms (ERs). ERs may be combined with wiring closets at any level, including main cross connects (MCs), intermediate cross connects (ICs), and telecommunications closets (TCs).

The space requirements for wiring closets are clearly defined in EIA/TIA 569. The minimum size for a TC or IC is 7 ft by 10 ft (2.2 m by 3 m). This space is designed to serve a 5000 sq ft area (500 sq. meters). A minimum size for an equipment room is 150 sq ft (14 sq meters). Table 11.2 shows the sizes for other served areas. Note that the metric conversions are greatly rounded in this standard. This is referred to as a "soft" conversion and you should use the dimension in the appropriate unit of measure for your normal building plan dimensions.

These sizes are really minimums, but should be adequate for most instances. In the case of equipment collocation with wiring closets, you will need to adjust the size of the room accordingly. Remember that you need to have space in front and back of racks, rails, and servers for routine maintenance in the wiring closet. While you may be tempted to put the recommended area into a narrow space, remember that your LAN workers, installers, and vendors will have to work in that area, and will be hampered by too small a space.

Power Requirements

Many wiring closets have networking equipment that will require electrical power. Power outlets for wiring closets should be on dedicated circuits. A dedicated circuit is one that serves only one outlet, or a series of outlets in one room. Standard practice in the U.S. is to equip each dedicated outlet location with a bright orange outlet (either a duplex or simplex receptacle) to distinguish it. The electrical circuit should be rated appropriately for the equipment that is to be connected. Some larger equipment may require its own separate dedicated outlet.

TABLE 11.2 Recommended Wiring Closet and Equipment Room Sizes

Served zone		Wiring closets	
Sq meters	(Sq feet)	Meters	(Feet)
500	(5000)	2.2 × 3	(7 × 10)
800	(8000)	2.8 × 3	(9 × 10)
1000	(10,000)	3 × 3.4	(10 × 11)

	Equipment rooms	
Workstations	Sq meters	(Sq feet)
Up to 100	14	(150)
101 to 400	37	(400)
401 to 800	74	(800)
801 to 1200	111	(1200)

You should be sure that an adequate number of electrical outlets are provided, based on the equipment that needs to be installed. Racks or banks of equipment may need an outlet strip or a power bar with multiple outlets. Allow lots of room for growth. A good rule of thumb is to provide twice the number of outlets you think you will need, and double that number! Remember to include open outlets that can be used to provide power to test equipment.

The standard recommends two dedicated duplex outlets AC outlets for equipment power, rated at 15 A 110 V (or appropriate ratings for your locale) on separate circuits. It also recommends convenience outlets wall mounted every 1.8 m (6 ft) around the telecommunications closet, in addition to any power requirements for collocation of an equipment room. Also specified is lots of ceiling mounted lighting, 540 lux (50 ft-candles). Equipment rooms should have enough HVAC capacity to maintain the room at 18 to 24 degrees Celsius (64 to 75 degrees Fahrenheit).

Another item to consider is power conditioning and backup power. Your power connections should, at the minimum, contain surge protection. In areas that are subject to power line fluctuations, you might want to consider ferro-resonant transformers that automatically compensate for sags and minor surges.

True power spikes must be dealt with by the use of line surge protectors. Several types of these exist, including gas-discharge protectors, semiconductor protectors, and metal-oxide varistors (MOVs). The MOVs are the most common and the cheapest, and provide the least protection—usually found in low-cost multioutlet strips. Their primary drawbacks in such an application are their slow response time and the user's inability to determine if they have lost protection from prior spikes.

A form of spike also occurs as a result of telecommunications wires that run outside your building. The most common occurrence results from a nearby lightning strike. The electrical discharge may induce destructive voltages in any outdoor cabling, including telephone and LAN cabling. It is very important to provide approved protectors on all telecommunications cables at the point of entry to your building. Fiber optic cabling with no metallic members can be an important tool in eliminating lightning risk.

Another potential power problem is a power outage. Equipment that might be disrupted by a power outage should have emergency backup power supplies provided. These power supplies usually connect between the equipment to be served and the AC power outlet. One type of uninterruptable power supply (UPS) is the backup UPS. When power is cut off, an inverter generates AC voltage from power stored in batteries. The generated power is switched to the connected equipment in a relatively short time period so that the equipment sees only a small break in the voltage. Another type of supply, the true uninterruptable power supply (true UPS), constantly generates power to the connected equipment. When there is normal line power available, the true UPS uses the AC line as its source of power. If there is a break in AC power, the UPS switches almost instantaneously to the battery source, with no appreciable interruption in voltage to the connected equipment. As you might expect, the true UPS is considerably more expensive than the backup UPS.

What type of equipment should be provided with emergency power? You should provide for equipment needed during an emergency, or equipment that can be harmed or disrupted by a power outage. Equipment needed in an emergency might include the telephone system, the alarm system, the video security system, or any network servers and hubs needed in an emergency. Some servers and other computer hardware are disrupted in a power outage and often must be manually restarted or "cleaned up" before normal use can resume. Emergency power, while temporary, may eliminate the restart process completely in a brief power outage. Some UPS systems alert servers to begin the shutdown process automatically, if it appears that the outage may be sustained. It also may be necessary to provide emergency power to network hubs to assist in the shutdown process. Unless user workstations are somehow provided with emergency power, it may be of little use to provide such power to the hubs and other networking equipment.

Backbones and Concentration

LAN backbones and cabling backbones unfortunately share a common name. During most of this book, we have been referring to backbone cabling rather than LAN backbones. However, in this section, we will actually be talking about the type of logical backbones that are used in LAN systems.

LAN backbones are often used to tie sections of the networking system together, so that workstations anywhere on the network can use resources, such as servers, that are not exactly on the same portion of the network. The workstations in a work area are connected to a hub-type device in the wiring closet. Hubs may be interconnected in a variety of ways, but often they are connected to a higher level of the network, called the backbone.

In a simple repeater backbone system, as shown in Fig. 11.3a, several stand-alone 10BaseT hubs are connected together via a 10Base2 coax segment. This setup has the advantages of 10BaseT star wiring to the workstations, but the disadvantages of coaxial bus cabling between the hubs. If any hub or coax connection fails, the entire backbone segment is likely to fail.

Figure 11.3b shows another alternative. Here, the hubs are connected to a higher level backbone hub, via 10BaseT. Because all hubs repeat the LAN signals to all ports, the servers can be located anywhere. Some variations of this scheme put each hub onto a circuit card that mounts in a hub chassis. The hub cards are interconnected at the backplane on a high-speed bus that is independent of the topology of the network. This bus serves as a type of backbone to all of the hub cards.

However, neither of these really tends to break up the network into smaller, more efficient chunks. Any network traffic is repeated throughout all of the connected hubs, effectively limiting the bandwidth of the entire network to a single 10 Mbps path (or 16 Mbps in the case of Token Ring).

a. A simple repeater backbone

Coax backbone

10 Base T Hubs

b. Multi-level hub backbone

c. A bridged backbone

Bridge or router   Bridge or router

d. A collapsed-backbone router

Multi-port router

Figure 11.3 Different types of backbones used in Ethernet networks.

Figure 11.3c shows several hubs connected via bridges (or routers) to a backbone LAN. Notice that servers on the local hubs are available to local workstations without transiting a bridge. This effectively makes the 10 Mbps bandwidth of Hub A add to the 10 Mbps bandwidth of Hub B, as long as the user on one hub has no need to use the server on the other hub. If a user on either hub needs to have access to a server on the backbone, they connect across the intervening bridge. Companywide applications are often placed on servers on the backbone to give access to any user without putting traffic on other parts of the network. Thus, a user on Hub A can use a backbone server at the same time that a user on Hub B uses a local server. Fig. 11.3*d* shows a variation using a multiport router.

This type of backbone hierarchy must be considered when planning the location of wiring closets, equipment rooms, and the cabling between them. Repeater backbones and bridge backbones can both be supported using the wiring standards we have covered in this book. However, the specific number of interconnecting cables between wiring closets may be dependent on the way you lay out your backbone system.

Hubs (or stacked hubs) are often placed in separate wiring closets. This means that the LAN backbone cabling must be run between closets to interconnect the hubs. These connections between wiring closets are traditional backbone cabling. The actual backbone bridges may be placed together in the same equipment room, or located individually at each hub that is to be connected to the backbone. In some LAN backbone topologies, a backbone ring must be formed linking each wiring closet. Token-Ring and FDDI are examples of this backbone ring method. Unlike the telephone scenario, the LAN backbone connections do not simply combine individual station cables into larger multipair cables. LAN backbones usually require one backbone connection for each group of hubbed workstations that are to be connected. Additional backbone cables may be run to provide backup for the primary cables or to provide closure of a ring. Backup cables or cables to complete a ring are an integral part of the backbone hierarchy.

By the way, nothing in this implies that a backbone is either fiber optic or copper, as either one may serve your needs well. Keep in mind that the allowable distances between closets in EIA/TIA 569 exceed the distances for copper wire 10BaseT or Token-Ring 16, set without reference to a specific LAN topology. However, the use of fiber optic cable can extend your cables to the limit of the TIA standard.

12

Testing
and Certification

The major reason to specify a standards-based LAN wiring scheme is to ensure that your installed cable will meet the needed level of performance for your intended use. To specify such demands three things. First, you must make a cable plant design that will meet the desired performance criteria when the other two requirements are met. Second, you must use cable and connecting components that are certified to the proper level of performance. Third, you must employ proper installation methods so that the performance potentials of the cabling and components are achieved.

This is a long way of saying that a well designed LAN wiring system, using the proper cable and connectors, and installed with good workmanship, will operate at the expected level of performance. Previous sections of this book have described all of these key concepts. This chapter will explore the ways in which you can verify the proper operation of your installed cable system.

We will discuss some of the ways in which cable and components are certified to a particular level of performance by their manufacturers. In addition, we will see how you can perform inspection and testing to ensure that your installed cable plant meets the required performance levels. Next, we will talk about the test equipment you will need to verify your installation. Finally, we will show how to use that test equipment to do cable certification.

Connecting Hardware and Cable Certifications

The primary standard for LAN performance certification in the US and many other countries is the *TIA 568-A 1995 Commercial Building Telecomunications Wiring Standard* (which we refer to by the shortened name TIA 568-A). An additional supplement, *TSB-67 1995 Transmission Performance Specifications for Field Testing of Unshielded Twisted-Pair Cabling Systems,* provides standards for the testing of installed links. Several other standards bodies, including the Canadian Standards Association (CSA) and the International Organization for Standards (ISO) have similar or coordinated standards, such as CSA-T529 and ISO/IEC IS-11801.

The TIA 568-A standard contains information regarding the performance levels of connecting hardware and cable used for telecommunications wiring. This standard contains explicit testing methods for some items and refers to documents from other organizations, such as ASTM D-4566, for other items. Cable and connecting hardware manufacturers offer components that are said to provide a certain level of performance, as delineated by the standard. The general measure of component performance is the "category" of operation. Categories 3, 4, and 5 are defined in the standard, along with their corresponding recommended uses. Please refer to Chapter 5 for an explanation of the five categories. The parameters that are specified in TIA 568-A pertain only to the performance of those components prior to their installation. The performance testing of installed link components is covered in the supplementary bulletin, TSB-67.

The actual testing of cabling components is the responsibility of the manufacturers of these items. Manufacturers may do their own testing to "certify" conformance with the standard, or they may use an outside testing facility. In some cases, manufacturers also have a recognized independent testing laboratory—such as UL—verify that the testing was performed properly. Some independent labs also offer a continuing verification or certification program so that manufacturers may further assure their prospective customers that their products meet the necessary performance criteria.

You are responsible for using only components of a proper category and for field testing the installed cable links to verify proper operation for the rated category. This process may be done by you, your installation contractor, or an independent third party to "certify" compliance at the rated category.

This means that if you expect to provide a LAN wiring system with a particular level of performance, you must use only components and cable that have been certified for that category of operation or higher. For example, for a Category 3 installation you may use either Category 3, 4, or 5 components, but you must use only Category 5 cable and parts for a Category 5 system.

A common misconception is that using Category 5 cable is always better than using Category 3 cable for an application that requires only Category 3 operation, such as 10BaseT. However, if you use components and installation techniques that are only appropriate for Category 3, you essentially get only that level of performance. The performance of a cabling system is only as good as its weakest link, and a category of performance is equal to the lowest certification of any component. You could well argue that a higher category of cable will ensure meeting a lower category of performance. The higher grade cable may improve performance margins a bit. It will, but the only gain to you is in those instances when a lower category of cable might reveal other marginal conditions, such as interfering electrical fields or longer-than-permitted cable runs. A much better reason for using a higher category component is that you could at a later time upgrade to Category 5 without repulling the wire. Be cautious, though, as this potential for later upgrade might be wasted if other Category 5 installation practices are not followed.

Performance levels

What does component certification really mean? Because recognized performance categories exist, a cable or connector that is marketed as meeting the requirements of a certain category should be tested to verify that claim. As the manufacturers are responsible for testing these components themselves, they often provide typical test results in their literature. Some manufacturers will provide detailed test reports upon request and offer product guarantees to back their claims.

In addition, independent laboratories offer a manufacturer verification testing programs that range from a one-time test report to a continuing test program with random testing. Obviously, you are more ensured of quality components with a continuing verification program than with a single test sample.

TIA 568-A outlines the performance specifications and very detailed test procedures that must be followed to certify components to a particular category. Annexes A, B, and C of TIA 568-A and ASTM D-4566 give the guidelines for component reliability testing and transmission testing for both UTP and STP cable types.

Pluggable connector components present some unusual challenges in reliability testing, as they must be checked for connectivity degradation. Random samples of each component are tested for degradation of contact resistance and insulation resistance in tests for durability, vibration, stress relaxation, thermal shock, and humidity/temperature cycle. This testing is to be done by the manufacturer to certify their parts to a recognized performance category. A cautious manufacturer might choose to have the certification verified by an independent testing laboratory, since a lot rides on the higher categories of operation.

Surprisingly, with regard to connector plugs and jacks, the TIA 568-A standard calls for only 200 insertion/withdrawal cycles without failure (usually defined by excessive contact resistance). Some manufacturers may specify a usable life of 500 to 2000 cycles for their connectors, although the standard only calls for 200 cycles. The point is made that these components have a finite life. Modular connectors that are used in patch panels with frequent use may be expected to exceed the standard's parameter over normal equipment life. Moreover, connectors that are used in test equipment, such as for cable certification, can exceed that number of cycles in a single afternoon! The user is only cautioned by the standards to "inspect the connectors for signs of wear." You would be much safer either to use connectors designed for high usage or to periodically replace the connectors on high-use equipment. You would also be wise to keep a usage log of your test cables and replace them regularly, as the modular plugs used on these cables have the worst degradation from multiple insertions.

Connectors must also be tested for transmission performance (attenuation and NEXT). Barring the degradation from excessive use, modular connectors have an innate contact resistance that is not significant when compared to the bulk resistance of a length of cable. The wire pins that form the connector are typically gold-plated to minimize this contact resistance. However, the modular-type connectors used in the current wiring standards do exhibit a significant amount of cross-coupling that appears as Near-End Crosstalk (NEXT). All of the 8-Pin modular style connectors are bad in this respect, but testing is done to make sure the NEXT is not worse than allowed.

Table 12.1 shows the attenuation and NEXT performance parameter limits for Category 3, 4, and 5 components. You can see from the values that little contribution to circuit loss is expected from the circuit hardware, as most comes from contact resistance. For comparison, the dc insertion resistance is only 0.3 ohms for a connector, compared to 9.2 ohms per 100 m for the cable. However, the NEXT contribution from the connectors can be significant. For example, at 100 MHz, the NEXT loss of the connecting hardware must be better than 40.1 dB, while the cable alone must be better than 32.3 dB for a 100 m length. The entire terminated and installed Basic Link allows only an additional

TABLE 12.1 Connecting Hardware Attenuation and NEXT Limits of TIA 568-A

	Connecting hardware limits (loss in dB)					
	Attenuation			NEXT Loss		
	Category			Category		
Frequency (MHz)	3	4	5	3	4	5
1.0	0.4	0.1	0.1	58.1	65.0	65.0
4.0	0.4	0.1	0.1	46.0	58.0	65.0
8.0	0.4	0.1	0.1	40.0	52.0	62.0
10.0	0.4	0.1	0.1	38.1	50.1	60.0
16.0	0.4	0.2	0.2	34.0	46.0	56.0
20.0		0.2	0.2		44.1	54.0
25.0			0.2			52.1
31.25			0.2			50.2
62.5			0.3			44.2
100.0			0.4			40.1

NOTE: The maximum attenuation values shown are specified at each frequency. Acceptable values of attenuation are a lesser (or equal) dB value than shown in the table. The NEXT Loss shown is calculated from the formula in TIA 568-A. The formula is based on assumed minimum values at 16 MHz of 34, 46, and 56 dB for Category 3, 4, and 5, respectively. Acceptable values of NEXT Loss are a greater (or equal) dB value than shown in the table, or calculated from the formula, at a given frequency.

3 dB (to 29.3 dB at 100 MHz) over the raw cable for all the intervening connecting hardware and installation anomalies. This is an area, by the way, where a higher "enhanced" grade cable can increase the margin of performance of a link. However, the connectors cannot be much improved and remain compatible with the standard modular design.

The low NEXT requirements of Category 4 and 5 are one of the primary reasons that the traditional screw-type wire terminals are not acceptable in those applications. You can easily observe the increased NEXT of screw-terminal jacks with any of the more sophisticated cable scanners. (An increase in NEXT is a lower dB number, as the measure indicates how low the crosstalk signal is with respect to a transmitted signal on another pair. You might think of the NEXT loss number as the amount of isolation from an interfering signal. A higher number is better.) Connector assemblies that are designed for the higher categories normally use an integral printed circuit board with careful routed wiring to minimize NEXT.

Screw terminals are also technically rejected on grounds of pair untwist, which must remain below 1.0 in for Category 4 and 0.5 in for Category 5. The station cable must often be untwisted in excess of those limits to reach the screw terminals. Screw terminal connectors also have an unacceptable length of untwisted wire between the screw terminal and the actual connector body. It is these factors that lead to the unacceptable level of NEXT and impedance anomaly in screw-type jacks.

You should be aware that there are some insulation-displacement (IDC) outlet connectors that use similar space wiring between the IDC block and the modular connector. Both 110 and 66-type mini blocks may be found in these types of outlet plates. These connector assemblies do not meet the needs for Category 5 (and probably not for Category 4) and should be avoided. This problem comes not from the connecting blocks but from the internal outlet wiring from the jack to the block. Do not confuse these jack plates with the acceptable style that integrates the 110 block into the connector as part of a printed wiring assembly.

Cable used in horizontal and backbone runs must also meet performance standards. Table 12.2 shows the attenuation and NEXT performance parameters for Category 3, 4, and 5 cable. Note that these levels are for the raw cable, not an installed link (see Testing Methods later in this chapter). Horizontal cable must meet essentially the same requirements as backbone cable. The attenuation and NEXT limits are calculated from a formula in the actual standard and the values in the table merely illustrate the calculation at specific discrete frequencies. Cable attenuation is known to vary with ambient temperature. The cable must also be tested (by the manufacturer) to meet adjusted loss limits at 40 and 60 degrees Celsius. The maximum allowable attenuation at these elevated temperatures for Category 4 and 5 cables is adjusted by a factor of 0.4% increase per degree Celsius.

Cable must also meet other transmission related specifications of dc resistance and unbalance, mutual capacitance, capacitance unbalance to ground, characteristic impedance, and structural return loss. The resistance and capacitance

TABLE 12.2 Cable Attenuation and NEXT Limits of TIA 568-A

	Horizontal and backbone cable limits (loss in dB)					
	Attenuation			NEXT Loss		
	Category			Category		
Frequency (MHz)	3	4	5	3	4	5
0.064	0.6	0.7	0.7			
0.150				52.7	68.7	74.7
0.256	1.2	1.2	1.1			
0.512	1.8	1.6	1.5			
0.772	2.2	1.9	1.8	43.0	58.0	64.0
1.0	2.6	2.2	2.0	41.3	56.3	62.3
4.0	5.6	4.3	4.1	32.3	47.3	53.3
8.0	8.5	6.2	5.8	27.8	42.8	48.8
10.0	9.7	6.9	6.5	26.8	41.3	47.3
16.0	13.1	8.9	8.2	23.3	38.3	44.3
20.0		10.0	9.3		36.8	42.8
25.0			10.4			41.3
31.25			11.7			39.9
62.5			17.0			35.4
100.0			22.0			32.3

NOTE: The maximum attenuation values shown are calculated from the formula in TIA 568-A. The attenuation value is per 100 m (238 ft) at 20 degrees Celsius. An increase of 0.4% per degree Celsius is added to the value shown for Category 4 and 5 cables. Acceptable values of attenuation are a lesser (or equal) dB value than shown in the table. Values for frequencies below 0.772 MHz are shown for reference only and are not required for testing.

The NEXT Loss shown is calculated from the formula in TIA 568-A. The formula is based on assumed minimum values at 0.772 MHz of 43, 58, and 64 dB for Category 3, 4, and 5, respectively. The NEXT Loss of the worst pair combination, for a length greater than or equal to 100 m (328 ft), is used. Acceptable values of NEXT Loss are a greater (or equal) dB value than shown in the table, or calculated from the formula, at a given frequency. The power-sum method of measurement is used for backbone cables.

components are primarily a function of the characteristics of the copper wire and the insulation material. These values are consistent and are easily described by a single set of values for each category, as shown in Table 12.3.

Characteristic impedance is specified as 100 ohms +/– 15% (or 85 to 115 ohms). However, it turns out that it is rather difficult to make a consistent measurement of characteristic impedance of twisted pair wire, due to variations in the cable's structure over its measured length. Consequently, the characteristic impedance value is derived by "smoothing" the results of measurement of a parameter called structural return loss (SRL). The SRL may vary rather wildly from 1 MHz to the highest referenced frequency (16, 20, or 100 MHz for Category 3, 4, or 5, respectively). For this reason, an absolute minimum limit for SRL is imposed for each category. When the SRL value is higher than the limit, it indicates that the impedance variation is better than required for the cable pair. The limit applies to the worst pair of the cable, so all pairs must be tested.

Because the twist pitch (lay length) of different pairs is varied to decrease NEXT coupling, each pair of a cable will show a variation in values of most of the measurement parameters. Some of the standard's limits are based on the worst pair's value, while others, such as attenuation, require all of the pairs to be at least as good as the limit value. This intentional difference in cable geometry also contributes to some of the variations we see in the SRL.

Patch cords and cross-connect jumper wires must meet the requirements of horizontal cable with some exceptions. It is assumed that the patch cords will use stranded wire for increased flexibility. Stranded wire pairs are expected to have slightly greater attenuation than solid copper, so an increase of 20% in attenuation is allowed over that for horizontal cable. The attenuation value is prorated for length, since patch cords are generally far shorter than horizontal cable runs. These wires need not be tested at elevated temperatures. In addition, the stranded wires must have a twist pitch (lay length) of 15 mm (0.6 in) or less.

The standard specifies a clear system of category marking for cable and connecting hardware. The category should be marked either with the words Category N (where N is the category number 3, 4, or 5) or with a large letter "C" with the category number in the center. These markings are shown in Fig. 12.1. The category of some wiring components can be very difficult to determine once they are installed. A sticker on a plastic bag, label on the package, or note in the data sheet is not of much use. The standard requires marking, and it should be easy for most manufacturers to stamp, emboss, or label each jack, plug, or cable with the rated category number. After all, you are paying extra for the performance. The best rule is, if it doesn't say Category 5 (or whatever), don't assume it is Category 5. Try to use marked components wherever possible.

TABLE 12.3 Cable Performance Requirements for Resistance, Capacitance, and Balance

Parameter	Applies to:	Conditions:	For category:	Requirement:	Mandatory?
dc resistance	Any conductor		All	</=9.38 ohms per 100 ohm	Yes
dc resistance unbalance	Any pair		All	</=5%	Yes
Mutual capacitance	Any pair	1 kHz	Cat 3	</=6.6 nF per 100 ohm	No
	Any pair	1 kHz	Cat 4 & 5	</=5.6 nF	No
Capacitance umbalance pair to ground	Any pair	1 kHz	All	</=330pf per 100 ohm	Yes

Note: All measurements are to be measured at or corrected to 20 degrees Celsius. The table applies to both horizontal and backbone cables.

SOURCE: TIA 568-A

Figure 12.1 Category of performance markings for components.

Pitfalls and specsmanship

The greatest danger in evaluating cable and connector hardware specifications is that the certifications may not be based upon the most current standards. The category nomenclature has evolved over a period of years from a loose general specification to a very specific (and tough) measured-parameter specification. Originally, the progressive ratings of performance were called Levels 1 through 5. Over time, the term "Category" was introduced and eventually adopted by several standards bodies, including the EIA/TIA, the UL, and the ISO/IEC. Now the performance parameters and testing methods have been further tightened, although the existing Category terminology remains. Participation in a continuous testing verification program by one of the major independent labs is the best assurance that the latest standards are used.

You may have cable or connectors on hand that do not strictly meet the current performance requirements of the categories. It is possible also that the testing was done using outdated procedures. You would be confident that the latest performance standards were met only if your hardware and cable were certified to meet TIA 568-A. A reference to SP-2840 would probably suffice, since it was the draft revision to the original EIA/TIA 568 that became TIA/EIA 568-A. In some cases, previously manufactured components may reference the technical bulletins that were issued prior to being incorporated, for the most part, into TIA 568-A. Test equipment should meet the TSB-67 requirements until they are incorporated into a future B-revision of TIA 568-A.

One pitfall you should avoid is the use of components that diverge from the recognized standards. For example, other styles of plugs and jacks may have much better NEXT performance than the ubiquitous 8-pin modular hardware. Typically, these other connectors and hardware are presented as having an advantage over the standard product. This is, you see, a way to say "standard" so that it seems ordinary. However, wiring standards are not only a set of performance parameters and installation practices, but also a standard style of interface to the user equipment that allows multivendor interoperability with a high level of performance. Any item that does not comply with the standards is an exception and may be potential problem for your network. This is, you see, also a way to hear "exceptional" that says it is noncompliant. For the time being, you may be taking a risk if you choose to use some of these better (but nonstandard) connectivity options.

What is specsmanship? *Specsmanship* is a term for the exaggeration or misleading presentation of product specifications that sometimes occurs in the marketplace. Fortunately, the existence of recognized national and international standards minimizes the occurrence of specsmanship, but it still occurs. Were there no standards, you can be sure that various manufacturers would make conflicting product claims that would be impossible for the buyer to resolve. The tough standards for telecommunications cabling make product differences clearly recognizable. All you have to do is know which standard of performance you want the product to meet, and then be sure that the products you use meet that standard.

There is a continuing controversy that involves connector performance. The controversy concerns the relatively high NEXT of all modular connectors and the measurable differences in longitudinal balance of some connector jack designs. Several types of nonmodular connectors are available that have significantly lower levels of NEXT than the popular 8-Pin modular design. Some of these connectors are used for field test instruments to reduce the contribution to NEXT from the test set and are quite acceptable for testing the Basic Link. However, a compatible lower-NEXT modular connector style would be much more desirable than attempting to replace more than a decade of installed modular connectors with another connector. Several manufacturers are working on the improved products, and we're already starting to see clever plug and jack designs emerge, but the longitudinal balance and resonance phenomena issues are more subtle. Resonance phenomena can cause short links (particularly under 15 meters) to exceed the allowable NEXT performance and is related to the return loss and/or balance of the link. These are subjects for further study by the TIA Task Group on UTP Link Performance.

Cable Plant Certification

In addition to using properly rated cable and components in your LAN wiring installation, you should plan to test your installed wiring system to the expected level of performance. Installing a cable plant or wiring system is a complex task, particularly if you intend for it to support high performance 100 Mbps networks. Even 10 Mbps Ethernet and 16 Mbps Token-Ring must have station cables that meet a certain level of performance to be considered reliable.

Most LAN wiring is put to use immediately after cable plant installation. Initial operation of the workstation is a good indication that the cable is all right, but it's not positive proof that the cable is in perfect condition—particularly if your workstations are currently using 10 or 16 Mbps technology, but you later plan to go to higher speeds. Also, it is not unusual for some of the station drops to remain unused until some later date. How will you know if those drops will work when you need them?

Would you have a light fixture installed and not try the switch? Probably not. You need to test each drop of your cabling system as well. The best way to test cable is to use proper test equipment that looks only at the cable performance. In that way, you avoid the nagging little network problems that obscure the real issues. Cable often gets blamed for problems that are really network hardware problems and bad software configurations.

The best time to do the testing is before furnishings and people have been moved in. At this point, all outlet jacks are accessible, without the need to move desks and other furniture. Any inspections and repairs that are needed can be done quickly and without disturbing anyone. Of course, modular furniture will have to be in place because the cable must be installed in the furniture modules before it can be tested.

The higher wiring categories in the current standards appear to offer no means to provide an intermediate point for interconnection to modular furniture. However, an optional (informative, not mandatory) annex to TIA 568-A describes a fixed multiple user telecommunications outlet that could serve 6 to 12 workstation areas (with two cables, each). This option suggests that the equipment cables (user cords) be routed through furniture raceways all the way back to the multiuser outlet. It may not be possible to do this without exceeding the recommended allowance of 3 meters (10 feet) for work area cables. If wiring closet jumper, patch, and equipment cords are kept short, it should be feasible to increase the work area cable length. The total length of these cables should be kept under 10 m, however. It is interesting that the standards allow a "transition point" to adapt round cable to flat under-carpet cable, but make no similar allowance for a modular furniture cable transition. A proposed TSB is expected to provide more details about specifications for multipair cables and multistation telecommunications outlets. See the online resources section of the Appendix for the latest information.

Testing installed cable

The testing of installed cable links is often referred to as "certification" of the installation. Certification is normally performed by the installer and involves testing each individual cable link to a recognized standard. The end result of the certification is a report that shows the actual measurements of each cable link and the pass or fail determination.

A dilemma initially existed in providing a method of testing and certifying installed telecommunications cable. The original EIA/TIA 568 standard specifically stated that the performance standards for cable and connecting hardware did not apply to installed systems. The reason for this was that it was known that an installed cable run had performance that was below that of cable tested at the factory. As a matter of fact, a correlation problem seemed to exist between cable tested with a factory RF network analyzer and with the field test instruments. Whether the field testers gave actual measurements of the parameters or judged the cable on pass/fail criteria, there were differences that might cause factory-certified cable to fail a field test.

Some of the differences were to be expected, as the installed cable had several wire terminations, cross connects, and modular connections that did not exist in the factory. Also, the standard factory test was (and is) to lay the cable out on a nonconducting surface (such as a concrete floor) or loosely coiled and use special test adapters to connect to a network analyzer. Cable in the walls is bent, pulled, tie-wrapped, run along metal pipes and steel beams, placed in conduit, and installed in all manner of nonideal conditions that affect its performance. You would expect it to perform differently. Moreover, there were no standards for accuracy for field testers and their performance varied considerably. No wonder there was no correlation.

Fortunately, the situation is much improved with the newer standards, such as TIA 568-A and TSB-67. The performance of installed cable is specified in detail, as is the accuracy of field testers. Even the required tests, testing methods, and method of reporting the results is specified. The correlation of field tester performance with laboratory network analyzers is also made. All of this lends a wonderful consistency and reproducibility to the testing of installed cabling that gives a real confidence factor to cable plant certification.

Certification for installed cables may be done two standard ways, Basic Link testing and Channel testing. Both of these testing methods are defined in TSB-67, which builds on the "worst case" link model contained in TIA 568-A Annex E. As we saw in Part I of this book, a Basic Link covers the horizontal cable from the workstation outlet at one end to the initial point of termination at a crossconnect block or patch panel in the wiring closet, including the test cords for the test equipment. The test connection at the wiring closet termination of a horizontal run may require special adapters unless the termination is directly into a modular patch panel. For example, the test cord would have to terminate in a 110-adapter plug if that type of connecting block were used.

A Channel includes the actual equipment cords (user cords, patch cords, and crossconnect wire) that connect to the network (or telephone) equipment at each end. Keep in mind that the word "actual" means just that. Once you certify a Channel with a particular set of equipment cables, those cables are frozen in place, as far as the standard's Channel testing method is concerned. If you change cables, to use a longer or shorter cable for example, you must recertify the Channel.

The Basic Link was designed to allow a cable installation to be certified once the horizontal cables were installed by the installation contractor. As a matter of fact, the link was originally called the Contractor's Link. Naturally, the user and patch cords were not in place at this stage and would have been impossible to test. Nevertheless, the performance of the entire link (the Channel) is ultimately what the user is dependent upon, so this test was also included as an important benchmark of overall link performance.

Which link test should you use? Basic Link certification is probably sufficient in most cases of new installations. It will ensure that the 90 m (maximum) of horizontal cable is performing to specifications. The Basic Link test is probably the most reasonable performance testing standard for a turnkey cable system installation, because the vendor may have little or no control over the cables used to connect to the equipment. Also, the build-out stage of construction is the best time to correct any problems that are revealed by a Basic Link test. After the network equipment is installed and connected, it will be easiest and most useful to do Channel testing. The Channel test is obviously very useful when dealing with network equipment vendors. You can rely on your cable system completely when it has been certified as a Channel.

Performance levels

Two types of performance levels should be considered when making field tests, the first being the performance category of the cable system. You should determine long before installation what category of performance you require and then test to that level. The second type of performance is the accuracy level of the field tester you intend to use to certify the cable.

We have discussed the cable performance categories at length in previous chapters. The three categories of operation specified by TIA 568-A and related standards are Category 3, 4, and 5. Other standards, such as some of the ISO documents, characterize similar performance levels, although with some differences. For example, the ISO/IEC IS-11801 specifies identical categories for uninstalled cable and components, but uses a class system for installed links. In most instances, the TIA categories are the same or tougher in test parameters and may be substituted with no problems. If you are in a situation that needs to reference these other standards, you should adjust your testing accordingly. While most of these standards on an international level are coordinated (also called *harmonized*) to use the same procedures, nomenclature, and limits, they may also contain deviations in language and some parameters that are appropriate in that region of the world. In addition, the standards bodies that promulgate these documents revise them on differing time cycles, and they may not be exactly synchronized to implement changes at the same time that the overall international community deems necessary.

TSB-67 identifies two performance levels for field testers, Accuracy Level I and II. Accuracy Level II is the toughest, and is always justified when you intend to operate your cable drops at or very near the length or performance limits, including user cords and jumpers. Table 12.4 shows the minimum requirements for the two levels. These accuracy levels are provided to ensure that the test instrument can actually measure the cable link parameters that need to be certified.

It is difficult to say which type of tester you need for certification. The potential problems actually occur only at or very near the acceptable performance limits. In a cable plant that uses very high quality cable and components and that has relatively short cable runs, you may never approach the limits required for your category of operation. However, if you use low-quality materials and questionable installation practices, you may see a lot of links at the performance limits. The Accuracy Level II is difficult to achieve, as its intent is to very closely emulate the performance of general-purpose laboratory analyzers costing 10 to 20 times as much. For that reason, it may be sufficient to get an Accuracy Level I tester or one that does not achieve Accuracy Level II on all parameters. Some testers may offer different performance levels on Basic Links and Channels, because the Channel test requires the use of the 8-pin modular connector at the test interface. As a LAN manager, you may choose to get a less expensive Accuracy Level I tester for your own casual troubleshooting use, and depend upon an installation contractor for Accuracy Level II tests when necessary.

TABLE 12.4 Field Tester Accuracy Levels of TSB-67

Performance parameter	Accuracy level I	Accuracy level II
NEXT accuracy:		
Basic Link	3.8 dB	1.6 dB
Channel	3.4 dB	1.5 dB
Attenuation accuracy:		
Basic Link	1.3 dB	1.0 dB
Channel	1.3 dB	1.0 dB
Dynamic accuracy	1.0 dB	0.75 dB
Length accuracy	+/ - 1m +/ - 4%	+/ - 1m +/ - 4%
Return loss	15 dB	15 dB
Random noise floor	50 - k* dB	65 - k* dB
Residual NEXT	40 - k* dB	55 - k* dB
Output signal balance	27 - kz* dB	37 - k* dB
Common mode rejection	27 - k* dB	37 - k* dB

Note: The constant k* is equal to 15 log (f/100) and thus varies with frequency. Performance and measurements are not required below 75 dB for Residual NEXT and for Random Noise Floor, and are not required below 60 dB for Output Signal Balance, Dynamic Accuracy, and Common Mode Rejection.

Testing costs

Proper testing costs money over and above the basic cost of installing and verifying the cable. As we will see in the next section, there are many tools that may be used to test LAN wiring. Some of these tools are fairly simple testers that check continuity and wire map (the proper pin connections). Use of such tools should be considered a normal part of cable installation and included in the price. These simple tests verify the dc electrical integrity of the link and find gross failures that would prevent the cable from functioning even in simple uses. This verification catches wiring errors and can find cases where the cable has been severely damaged after installation. Verification, by visual inspection and continuity check, once was considered sufficient testing for telecommunications cables. This is no longer the case.

The demanding cable performance that is required by modern LAN applications can no longer be met by such simple testing methods. To truly test a LAN wiring system to meet one of the more rigorous performance categories requires the use of very sophisticated test equipment. Specialized field testers have been developed to reduce the cost of testing. These testers are far less expensive to purchase than the ultrasophisticated laboratory RF network analyzers (these test all sorts of electronic networks, not local-area computer networks). Yet, the field testers yield useful results that closely match the measurements to the LAN category standards, produce quick pass/fail indications, and print full reports. These field testers are also very easy to use. They can run completely automated tests, store the results, and pinpoint problem locations on the cable under test. Most of the testers will operate for several hours on battery power, so you don't even have to find a power outlet.

Although these testers provide many advantages, they are still much more expensive than the simple wiring verification tools, and they require additional skill to operate properly. Consequently, you should expect to pay extra for having your cable"scanned" with one of these field testers. The cost of the tester, the skilled operator time, the preparation of reports, and the possible retesting should be considered when budgeting for the additional cable scanning step. You should allow a budgetary cost from 5 to 15% of the installation cost per drop to certify to Basic Link standards. The cost variation is to allow for special circumstances, such as an unusually difficult and expensive per-drop cost. If the scanning is done by a separate contractor, you should allow an additional factor for problem determination or visual inspection. Scanning is best done before personnel have moved into the office space. Although it involves only a brief visit to each workstation outlet, labor costs will be higher if furniture (and people) has to be moved.

As we have said before, the high testing standards are necessary for your wiring system to work to the limits of the standards. If you intend to run only 10BaseT or Token-Ring on Category 3 or higher wire to a small number of workstations and your drop lengths are much shorter than the maximum length, you may not need to run a complete certification scan at all; simpler visual inspection and continuity verification may be all that is needed. It will certainly reduce the cost of your installation, and there is little chance that such a small network will not work for those applications. After all, we have been running twisted-pair 10 Mbps networking on plain old telephone wire for years. For the most part, the problems only arise when the runs are longer than about 50% of the maximum TIA length (that would be about 45 m or 150 ft).

If your intention is to install a full Category 5 wiring system, however, then you need to plan for a complete certification scan. The component standards and installation practices are simply too tough to not test completely for proper performance. If you have an existing facility that is supposed to meet Category 5 and you have not previously tested it to 100 Mbps, then a new scan would be appropriate. Please keep in mind that the testing standards for installed cabling were only released in 1995, so installations prior to that time may have used Category 5 components and accepted practices but could not have been tested to the current standards. Prior to the release of the testing procedures and limits in TSB-67, standard practice was to use Category 5 cable and components, install using the best current practices, and test using whatever cable scanners were available at that time. The older EIA/TIA 568 standard specifically excluded installed cable and connectors from its scope. The new TIA 568-A states that such items are "under study" and defines a worst case link; the actual installed tests are in TSB-67. At some point, the requirements of TSB-67 will be incorporated in a B-revision of TIA 568-A, but until that time, you may want to reference both documents in a cable plant specification.

Test Equipment

The proper use of test equipment is one of the most important aspects of a successful LAN wiring installation. Test equipment may be subdivided into two types: basic troubleshooting equipment and diagnostic/measurement equipment. Basic troubleshooting equipment consists of devices that check continuity, dc resistance, and simple wire maps. Diagnostic/measurement equipment is used to measure cable parameters, determine compliance with standards, and diagnose specific cable problems.

Which type of test equipment should you have? The answer to that question depends on the type of work you need to do. If you do any amount of cable installation or troubleshooting, you need most of the basic test equipment items described below. If you need to test or certify cable to cabling standards, you will need a field tester capable of meeting the accuracy requirements of the category you are testing. Fiber optic cable requires special test equipment, even for routine tests. For a fiber optic installation, you should have fiber attenuation measurement equipment, at the least, and you may need more sophisticated scanning equipment to locate fiber faults.

How much money should you spend for test equipment if you are the network manager? A good rule of thumb is to expect to spend 2-5% of the value of your networking equipment (including workstations and file servers) for test, analysis, and monitoring equipment. As an alternative, you could allow 10% of the value of just the cable, hubs, and bridges for test equipment. The point is that you need to have a reasonable capability to test and troubleshoot your network. Without the proper equipment, you are flying blind. You will suffer from periodic cases of "equipmentus substitutis" (as evidenced by lots of running around while carrying computers, hubs, and cables). To avoid this syndrome, simply obtain the necessary test equipment to simplify and speed your troubleshooting. You also need a certain amount of test and monitoring equipment to verify the health of your network and to spot trends that may have a negative effect on performance in the future.

The rest of this chapter is devoted to the essential test equipment you will need and the methods to use in employing the more sophisticated testers. In addition to the physical layer equipment we show here, you may also have need for LAN protocol analyzers that deal with the higher layers of networking. These protocol tools will help you answer the age-old question, "OK, if it's not the cable, then what is it?"

Continuity testers (voltmeter, continuity tester)

The most basic item of test equipment that should be in your toolbox is a good electronic voltmeter. This voltmeter should be able to read dc volts, ac volts, and ohms. It should be durable and have an easy-to-read display. A "hold" function is also nice, as it allows you to make a measurement and later read the display. This is great for those times when you are deep inside a punchdown block or in an area with low light. Many voltmeters also have a continuity function with an audible beep (some of us consider this a mandatory requirement).

Voltmeters can make quite a few basic tests. For example, you can check continuity and dc resistance on a new reel of cable. If you know the resistance of a full reel, you can easily estimate the remaining length from the dc resistance value. With test adapters, you can check continuity of a jack, patch panel, or patch cord. You can make a few simple adapters, wrap-around plugs, and break-out cables to help you use the meter for continuity and voltage measurements. You can also check for potentially dangerous voltages that may exist on cables run between wiring closets or between buildings.

A type of voltmeter that is particularly handy in the field is the probe-style digital voltmeter. This type of meter, shown in Fig. 12.2, has the entire meter circuitry and display in one oversized probe. The other probe is connected via a short wire to the probe/meter. With this type of voltmeter, you do not have to find a flat surface for the meter—it is right in your hand.

Beeper and LED continuity testers are also available and may be effectively used to test wires and connections. Continuity testing merely verifies that an electrical contact has been made, and does not really measure how good that contact is. However, it is rare in these systems to get an "ohmic" contact (one that has a significant resistivity). Since IDC connectors are used almost exclusively these days in LAN wiring, the metallic wire is scraped clean of contaminants during the termination process and the actual point of contact remains virtually gas-free, eliminating oxidation of the contacting surfaces that would lead to increased resistance. Dusty and corrosive environments are an exception to this rule.

Cable wire map testers

A very good device to test an installed cable link is the wire map tester. These little testers give an indication of cable continuity and proper pin-out. They are generally used for installed cables or assembled cords. One type has a main tester and a remote unit. The two units communicate across the cable to perform a wire map test and, of course, a basic continuity test. Most of these testers can easily find all of the standard connection errors and problems, such as opens, shorts, reversed wires, and crossed connections. (These cable faults are all explained in the next section.) Simple miswirings are probably the most common cable installation problem, and the easiest to fix.

The simple wire map tester performs a simple dc continuity test and checks for undesired connections. It does not check for proper pairing, as with a split-pair fault. A *split pair* is a condition where one of the two wires in one pair is accidentally exchanged for one of the wires in another pair. For example, the white-green and the white-brown might be exchanged. This actually means that two pairs are mixed and the balance and self-shielding properties of the paired wires is lost. In this type of miswire, dc connectivity tests fine and it would pass a simple wire map test, but the cable would fail if tested for ac signal balance by a cable scanner. In many instances, a cable with a split-pair

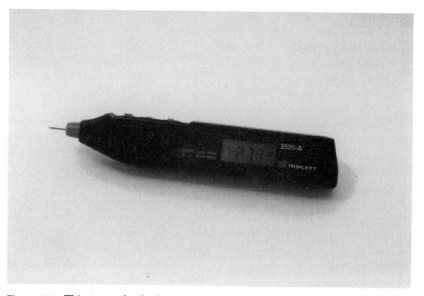

Figure 12.2 This type of volt-ohm meter has the digital display in one oversized probe.

fault will cause a LAN link to fail, which is why it is important to test for this with the more sophisticated scanners.

A specialized version of the wire map tester is used to test assembled cables. This cable continuity tester has two modular jacks where each end of the cable is plugged in. When a button on the tester is pressed, LEDs light to indicate continuity of each wire in the cable, one by one. A Pass is indicated by observing that the LEDs each light up in order through the number of pins in the cable.

This type of tester may be used for 6-pin modular connectors in addition to the 8-pin connectors used for LAN wiring. Frequently the same test jack is used for both sizes of connectors. This creates a potential false-failure problem for pins 1 and 8 of the 8-pin plug. Although the 6-wire plug will fit into the jack of the tester, the plastic sides of the plug may permanently bend the jack's first and eighth pins, making them no longer contact Pin 1 and 8 of an 8-position plug and giving a false indication of failure. The solution is to never use 6-pin plugs in a tester you intend to use for 8-pin plugs! If you believe a false failure has occurred, you can usually form a small stiff wire (such as a paper clip) to bend the jack wires back into position. Then test the cable again.

Cable tracers

How do you find a single cable in a bundle of cables? How can you quickly verify that an outlet is properly marked? The answer is to simply use a high-impedance tone generator and an inductive cable tracer, shown in Fig. 12.3. These are two of the handiest tools around, and no serious cable installer should be without them.

The tone generator was originally developed by the telephone industry to assist its installers in tracing wires into and out of the distribution frame. Any unmarked jack could quickly be traced by placing a "tone" on the jack and using the telephone test set (commonly called a *butt-in set* or just *butt set*) to find the tone. The tone's distinct warbling sound could be easily distinguished from other sounds on the telecommunications wiring. However, the butt-in set had to be physically connected to one of the cable's wires to hear the tone. Something else was needed.

That something is called an inductive pickup. This device uses a coil and a probe point to magnetically couple to a cable so that you can pick up the tone without having to make electrical contact with any of the wires. There are two types of inductive cable tracers: the passive pickup and the amplified pickup. The passive pickup contains only a pickup coil. Often called a "banana probe" because of its shape, this tracer requires the use of a butt-in set to hear the tone. The amplified probe contains a battery-powered amplifier in addition to the pickup coil. A momentary contact switch conserves battery power when not in use. The amplified probe can be quite sensitive, as long as the area is electrically noise-free. However, certain cable areas may generate too much electrical noise for the amplified probe to function effectively. This is particularly true of "live" telephone circuits, where the residual tone will be faint at best. Many installers carry both types of probes to the job.

Figure 12.3 A cable tracing tone generator and inductive amplifier. *(Courtesy of progressive Electronics, Inc.)*

Cable scanners

The advent of structured wiring standards has increased the need for very sophisticated portable field test sets that thoroughly test a cable run. These test sets are often called *scanners* because they use time-domain reflectometry (TDR) techniques to scan along the cable's length for anomalies. Reflectometry is somewhat akin to a radar for cable. A pulse is sent along a cable pair and then the cable pair is monitored to see what comes back (and precisely when it does). The type of reflection that returns is related to whether there is a short, an open, a termination, or a deformity along the cable. Of course, a cable may have several conditions that reflect part of the pulse. The length of time that it takes for a pulse reflection to return is directly related to the distance from the test set. A simple formula defines the distance to the reflection:

$$d = tkc/2$$

where

d = distance

k = ratio of the medium's nominal velocity of propagation (NVP) to the speed of light, expressed as a decimal fraction.

c = speed of light (units/second)

t = time (seconds)

The units of the distance will be the same as the distance units of c. In other words, if c is in meters per second, distance will be in meters. The nominal velocity of propagation (NVP) is the decrease in actual propagation of the pulse from the ideal free-space speed of light, when traveling through the cable. The NVP is alternately quoted as a decimal fraction or as a percentage. It ranges from about $0.6c$ to $0.9c$ for electronic cable. Dividing by 2 is required because the time interval is a round trip, but we are only interested in the one-way distance.

Fortunately, all of these portable scanners are computerized, so they make the calculations for us. Typical cable scanners are shown in Fig. 12.4. Scanners can automatically check the wire map for wiring faults, measure NEXT and attenuation, and check the cable impedance and return loss. Cable scanners also check for wire length, to ensure it is below the appropriate limits.

The TIA 568-A and TSB-67 standards specify the types of tests that must be made, the way in which the reporting must be done, and the accuracy levels of these field test sets. Accuracy levels and reporting requirements were explained earlier in this chapter. A description of testing methods is in a later section of this chapter.

Many cable scanners have an additional LAN traffic function that is very useful to the LAN manager. This function allows the passive monitoring of a LAN connection to measure traffic and gather some basic LAN statistics. Most of these testers, for example, can determine LAN utilization (expressed as a percentage). Some also measure other LAN traffic parameters and show potential error conditions such as collisions. Even though this goes beyond the cable testing and certifications for a generic cable system, it can be very useful feature to monitor the basic network functions.

TDRs/OTDRs

Although cable scanners use a form of time-domain reflectometry to perform the important certification measurements, a traditional TDR may provide additional insights into cable operation. Expensive laboratory TDRs may not be needed, as some of the field testers mentioned in the previous section now incorporate TDR graphics in their instruments. The actual graph of a cable's reflections can reveal many things about the handling and installation of the LAN wiring that simple Pass/Fail reports and raw measurements obscure. A typical TDR screen is shown in Fig. 12.5.

Optical fiber has a more sophisticated testing requirement than does metallic cabling. The same low-loss characteristics that make fiber optics important in transmitting longer distances make loss measurements critical to ensuring fiber cable performance. Optical cable is also subject to impairments from bending and stress that cause additional signal loss. It is often much more

Figure 12.4 Typical cable scanners for Category 5 testing. *(Courtesy of Datacom Technologies, Inc.)*

Figure 12.4 *(continued) (Courtesy of MICROTEST.)*

expensive to replace an entire run of fiber than simply to find the location of the fault and repair it by splicing the cable, particularly if outside plant cable is involved. The problem is how to find the fault.

An optical version of the TDR, called an optical time-domain reflectometer (OTDR) is the solution. This device operates in the same manner as the conventional TDR but uses a pulse of light, rather than an electrical pulse. Modern OTDRs display a graphic that shows the signal return of the fiber along its length. By moving a cursor, you can pinpoint the distance from the source to any anomaly. Then the area of the fault can be located by direct measurement. If the fault is obvious from the outside of the cable, the bad fiber may simply be cut and spliced, via fusion or coupling. Otherwise, it may be necessary to replace a section of cable, similarly spliced.

OTDRs may provide built-in connections for both single mode and multimode fiber. Since the TIA 568-A standard recognizes both fiber types, both will need to be tested in large systems. The standard also specifies that the same type of fiber optic connector, the 568SC, be used with both fiber types. Single mode fiber requires special laser-type sources to couple light effectively into the fiber. Since part of the measurement involves optical loss, it is essential that the coupling be to the proper type of source. Adaptation to different styles of fiber optic connectors may be required, as the 568SC connector is not yet in wide use.

Basic fiber optic testers

Optical fiber may also be effectively tested with simpler equipment. Very basic fiber test equipment couples light into one end of a fiber link and detects it at the other end. Some of these testers use visible light, so you can theoretically see the "light at the end of the tunnel." However, some higher power links may have enough infrared light output that it could be dangerous to the eye. Since the infrared light is invisible, you should never look directly into a fiber from

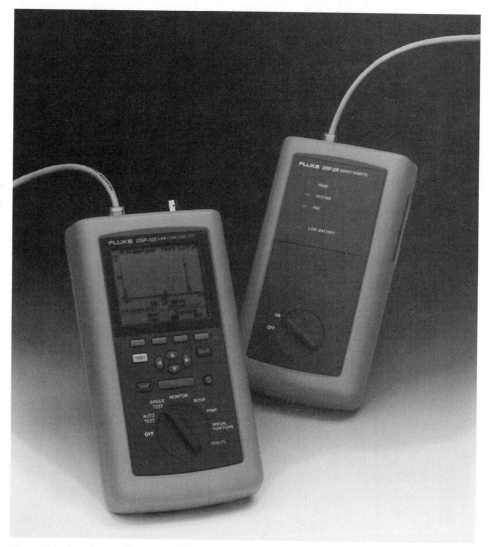

Figure 12.4 *(continued) (Courtesy of Fluke Corporation.)*

Figure 12.5 A time-dome reflectometer displays discontinuities and anomolies. (*Courtesy of Tekronix.*)

any close distance. A good practice with one of these testers is to use a detector or to observe the light, if visible, on another surface, such as a sheet of paper.

Optical loss measurement may also be used to test optical fiber. This equipment measures simple optical loss over a fiber link by placing a calibrated light source at one end of the cable and a sensitive optical power meter at the other end.

Testing Methods

The Channel and Basic Link

The installed cable link components of a horizontal link are measured either as a Basic Link or as a Channel. The difference between the two is that the Channel includes the actual equipment cords (user cords), patch cords, horizontal crossconnect, and crossconnect jumpers that will ultimately connect to the LAN equipment at each end of the horizontal link. The Basic Link includes two test equipment cords, one at each end (excluding the connectors at the tester), while the cords at the ends of the Channel must be plugged directly into the test instrument interface with no allowance for test cords. The Basic Link and Channel are shown in Fig. 12.6.

As we mentioned earlier, the reason for the two types of links is to facilitate testing at both the construction phase and the post installation phase. It is unusual for network hubs and workstations to be available during the construction phase of an installation. It would be time-consuming and difficult to wait to detect and repair bad cable runs at some later time. Moreover, the cable installer may have little responsibility or influence for the equipment cords. This two-step method of cable link specification allows the installed "in-the-wall" cable to be tested as soon as installation is completed. The complete link, from network hub to workstation, can be further tested during installation of that equipment.

a. The Basic Link

b. The Channel

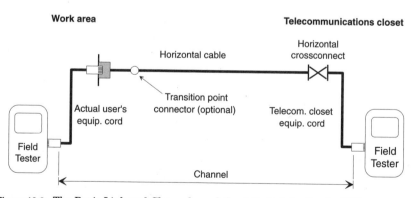

Figure 12.6 The Basic Link and Channel are defined for link testing by TSB-67.

Equipment & testing requirements

The purpose of the standards in TSB-67 are two-fold. The full title of the TSB is "Transmission Performance Specifications for Field Testing of Unshielded Twisted-Pair Cabling Systems." One explicit purpose, then, is to define cabling system performance after the cable has been installed. The other purpose, which is implied by wanting to test in the field, is to define field tester performance. Early field testers did not always produce consistent results when compared to ultra expensive laboratory testing methods. So, the standard states which tests will be done, what measured performance levels are important, and what tester accuracy and reporting is required.

The testing that is required includes parameter measurement for the following tests:

- Wire map
- Length
- Attenuation
- Near-end Crosstalk (NEXT) Loss

As more is learned about the requirements of networks operating at the higher levels, additional tests may be added to the field test specifications, and others may be modified. For example, you will notice that impedance and structural return loss (SRL) that are measured on uninstalled cable (per TIA 568-A) are missing from this specification for installed links. This is not because it is considered unimportant, but because it is not yet known (or agreed) what levels are appropriate. On the other hand, the cable length, attenuation, and NEXT are currently tested. The length of the cable run is important in influencing attenuation and NEXT values, but perhaps the ratio (attenuation to crosstalk ration, ACR) is a better indicator of overall system performance. Likewise, very high speed networks may care as much about the propagation delay as the physical length of the cable, yet transmission delay is not yet specified nor tested. These new tests could be a boon to transmission engineers but a bane to cable system designers and installers. Can you just imagine telling your installer to keep cable runs under 300 nanoseconds? Where do you get a tape measure that size?

Let's look individually at each of the tests that are now required.

Wire map

The wire map tests checks for wiring errors and any significant cable faults, such as shorts or opens. It is a very good idea to check the reeled cable for shorts or opens before installation, as it will be very hard to prove that the defect was not caused during installation. You can also perform most of the other tests, as the reeled cable should easily pass them before it has been installed. Keep in mind that, if you want to check the NEXT, you will need to use a properly rated connector because its NEXT will often be significant compared to that of the cable.

The wire map test will check for:

- Continuity
- Shorts
- Crossed pairs
- Reversed pairs
- Split Pairs

Figure 12.7 shows the correct pinout/pairing and most of the common wiring errors. In the standard TIA 568-A wiring scheme, the wires are paired 1-2, 3-6, 4-5, and 7-8. The pair numbering, 1 through 4, varies slightly between the T568A and B wiring patterns. Color codes for the cable are assigned on a pair basis, so (as an example) Pair 2 will always be Orange and White but may be assigned to the 3-6 pins or to the 1-2 pins, depending upon wiring pattern. The tester has absolutely no idea what color the wires are! The technician will have to visually inspect them if the design requires that a particular pattern be used. However, if both ends of a cable or run are wired with the same pattern, the field tester will pass the cable. If not, it will show a crossed pair condition, as in Fig. 12.7c. The field tester also easily finds split pairs, shown in Fig. 12.7d, an impossible task for a dc continuity tester.

Length

Length measurements are made on two types of cable links, the Basic Link and the Channel. Recall that these two links differ in that the Channel includes the actual user equipment cables that will be attached to the hub, the workstation, or whatever. For the field tester, this means that the Basic Link must include the test equipment cords at each end, but the Channel must terminate directly into the tester (or somehow exclude the effect of test cables from the measurement).

The TIA 568-A standard requires a horizontal link to be 90 m (295 ft) or less, with an allowance for an additional 10 m (33 ft) for all user cords, jumpers, and patch cords. Because this horizontal link length cannot directly be measured electrically by the field tester, which must somehow connect to the link under test, an allowance is made for two 2 m test equipment cords. This means that the Basic Link is defined to have an acceptable length of 94 m. The Channel, which includes the user cords, can be as much as 100 m. From a practical standpoint, we would recommend that you use test cords that are exactly 2 m in length. If you use a shorter cord, the tester may pass a link that exceeds the limit. If you use too long a cord, you may reject links that actually are legal. Some testers require a test cord of at least 1m. Never fear, because of individual cable variations in NVP, the TSB allows an additional 10% before it declares a length failure (plus a factor allowing for the +/- 1m +/- 4% tester accuracy, although pass/fail results in this zone must be marked with an asterisk).

An interesting distinction should be pointed out in regard to length. The physical length of a Channel or Basic Link is an important parameter. Physical

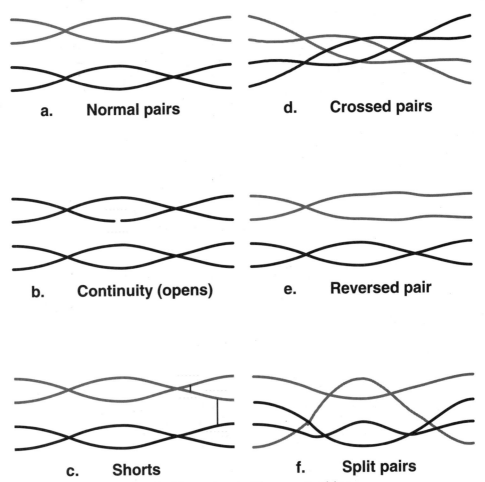

a.   **Normal pairs**

d.   **Crossed pairs**

b.   **Continuity (opens)**

e.   **Reversed pair**

c.   **Shorts**

f.   **Split pairs**

Figure 12.7   The correct pinout/pairing and most of the common wiring.

length may be measured or determined from markings on the cable. Cable testers estimate electrical length from the propagation delay of the cable. Because propagation delay depends upon the physical characteristics, such as the twist pitch (lay length) and dielectric properties of each pair, it varies with different categories of cable and even with different lots of cable from the same manufacturer. To accurately correct for manufacturing variations in cable, it is very important to measure the propagation delay of each reel of cable installed. The cable pair with the shortest propagation delay is used.

The procedure for calibrating length on a field tester is to physically measure a length of cable from the reel to be installed and calibrate the tester through an electrical length measurement on that same sample of cable. The particular length of cable you use is not important, just that its exact length be known. However, accuracy of the calibration is increased with a longer piece of cable. A length of 15 m (or about 50 ft) or longer is commonly recommended. Realistically, it would not be practical to cut and measure every reel of cable at a large job, but you should calibrate for each different cable lot if possible. You might cut a length for one of your cable runs to use to calibrate the cable, so as to not waste the sample.

Testers are required to have a length range of at least 310m, so that a 1000 ft reel of cable can be measured. However, until the cable on a particular reel is calibrated for NVP, this measurement should be considered only approximate. Stated lengths on reeled cable are not exact and should not be used exclusively to calibrate the nominal velocity of propagation (NVP) of a cable.

Attenuation

The maximum attenuation for the purposes of testing the Basic Link or Channel has been based on the attenuation values given in TIA 568-A for horizontal cable, connecting hardware, and jumper or patch cords. It prorates the loss values, given in dB per 100m, and applies them to the worst case length of the link. Table 12.5 shows a summary of the allowable attenuation values for the Basic Link and the Channel at selected frequencies. TSB-67 contains the formulas for calculating each of the components of Basic Link or Channel attenuation. The actual measurements should be made at intervals of 1 MHz or less in the appropriate frequency band, but the tester will take care of this automatically.

TABLE 12.5 Attenuation Limits for Installed Cable Links, in dB.

Frequency	Basic Link (dB)			Channel (dB)		
	Cat. 3	Cat. 4	Cat. 5	Cat. 3	Cat. 4	Cat. 5
1.0 (MHz)	3.2	2.2	2.1	4.2	2.6	2.5
4.0	6.1	4.3	4.0	7.3	4.8	4.5
8.0	8.8	6.0	5.7	10.2	6.7	6.3
10.0	10.0	6.8	6.3	11.5	7.5	7.0
16.0	13.2	8.8	8.2	14.9	9.9	9.2
20.0		9.9	9.2		11.0	10.3
25.0			10.3			11.4
31.25			11.5			12.8
62.5			16.7			18.5
100.0			21.6			24.0

NOTE: The Basic Link (sometimes referred to as the contractor's link) is comprised of the telecommunications outlet in the work area, the horizontal cable (which may include one transition, such as to under-carpet cable), the termination in the wiring closet, and, necessarily, the test equipment cords at either end. The Channel is comprised of the components of the Basic Link, excluding the test equipment cords, plus the crossconnect field or patch field to the user equipment in the wiring closet, and the user cords at either end (which are connected to the test equipment in place of test cords). The test equipment connection is excluded in either case.

SOURCE: TSB-67

The attenuation test limit is for a worst case link. The actual attenuation of a measured link will be less is the cable run is shorter than allowed. However, in most testers, the measured loss is compared only to the limit for worst case. This means that the test will not directly catch cable that has greater than expected loss (based on the uninstalled cable limits of TIA 568-A), but the extra attenuation per meter should have no effect on link performance. If you wish to catch substandard cable, you will have to compare the actual loss measurement with the expected loss for that particular length of cable.

As with the specification for uninstalled horizontal cable, there is an allowance for the effect of temperature on the attenuation of the cable. The allowance is 1.5% (of the value in dB) per degree Celsius for Category 3 and 0.4% for Categories 4 and 5. To illustrate this, consider a cable installed in an attic, where the actual temperature is 30 degrees Celsius, rather than the nominal 20 degrees of the standard. The attenuation limit for a Basic Link at 10 MHz would be raised from 10.0 dB to 11.5 dB (an increase of 10.0 dB × 1.5%/degree × 10 degrees) for Category 3, but only to 10.4 dB (10.0 dB × 0.4%/degree × 10 degrees) for Category 4 or 5.

Two other environmental factors are known to cause changes in attenuation, metallic conduit and humidity. The difference expected for cable in conduit is up to 3% increase, but the standard allows no latitude in test limits. It should be expected that long runs in close proximity to other metallic objects might also cause an increase in attenuation. You should keep this in mind in planning your design. The humidity effect is a new consideration. Some cable manufacturers have stated that the humidity effect is cumulative over time in PVC or plenum-rated PVC-compound jacketed cable and may cause cable that originally tested good to fail after a period of exposure to high humidity. The standard makes no allowance for the effects of humidity; the cable link must pass the attenuation limits regardless of the humidity. This might be an argument against using the less expensive PVC insulation or plenum-rated PVC-compound jacket and might also be a warning signal for some of the new copolymers that contain other thermoplastics mixed with the more expensive TFE compounds.

Near-end crosstalk (NEXT) loss

The theory behind twisted-pair wiring is that the twisting rotates the magnetic fields from the wire with the twist and coupling to nearby pairs or other objects is minimized. However, some coupling does exist. How much coupling occurs between pairs of wires on the same cable is of particular interest. If too much signal is coupled from, say, the transmit pair to the receive pair of a connection, the receiver will not be able to distinguish the far-end signal from its own transmissions. This coupling is illustrated in Fig. 12.8.

Coupling between pairs of a cable is called *crosstalk*. Because the crosstalk between transmit and receive pairs is greatest when the transmitted signal is unaffected by attenuation, the worst case occurs at the same end of the cable where a test signal is transmitted. This is called the *near end* and yields the expression Near End Crosstalk, or NEXT.

Figure 12.8 The NEXT measurement. If too much signal is coupled from the transmit pair to the receive pair of a connection, the receiver will not be able to distinguish the far-end signal from its own transmission.

NEXT is measured in dB below the test signal. Technically, NEXT would be a negative dB number, but we turn it around by calling it NEXT Loss. This means that a large NEXT number in dB is considered low NEXT, as the amount of crosstalk is much farther below the test signal. A high NEXT would be a smaller number in dB, indicating much less isolation from pair to pair. Very logical, n'est-ce pas?

NEXT must be measured in both directions on the cable link, to prevent the situation where one end of a cable passes NEXT and the other end fails. Single-ended measurements are often made in the wiring closet, as it eliminates carrying the tester around to each workstation outlet. Excessive untwist at the connector is more often found at the workstation end, however, because the outlets naturally separate the conductors, and because the individual jack plates are more difficult to inspect. Also, the workmanship may be less because the supervisory control may not be as great as would exist in the wiring closet. So if you only measure at one end, you are likely to miss more of the high NEXT connections. The equipment at both ends require appropriately low NEXT to operate properly.

A related figure of merit for a cable link is Attenuation to Crosstalk Ratio (ACR). Although not yet specified by the standards, ACR is a very good measure of how the attenuation and the NEXT can influence the performance of networking interface devices. Attenuation may be used to calculate the received signal strength of a signal transmitted from the far end. NEXT can yield the level of the interfering crosstalk signal from the near-end transmitter. The near-end receiver is thus required to cope with a signal-to-noise ratio that is basically the ratio of the two signals, the attenuated far-end transmission and the near-end crosstalk. Since both attenuation and NEXT are directly measured as part of the testing, it is very simple to ratio the two to produce the ACR figure. The higher the ACR number, the better the expected performance. Since a ratio of two logarithmic quantities is their numerical difference, the ratio is simply:

ACR = |attenuation(in dB) – NEXT Loss (in dB)|

One of the things that concerns network hardware designers is that the ACR decreases steadily with frequency. This means that at 100 MHz, the ACR margin may be very small indeed. Thus, it may be of benefit to those planning to eventually go beyond 100 MHz to purchase cable with the highest ACR possible.

Reporting pass and fail criteria

An important part of TSB-67 is the carefully detailed requirements for the reporting of test results. In general, the standard requires that each measured result be reported in units appropriate for the test and that a Pass or Fail determination be reported. Measurements, such as for attenuation or NEXT, that might be very close to the limits are reported with an asterisk in addition to the Pass or Fail.

A data report is simply the visual indication of the field test set. In all of the modern cable scanners, the visual display is a liquid crystal display (LCD) that also allows the display of some graphics. Of course, it is important for a permanent record of the report to be made, so the testers also allow the test data for each cable to be stored and later printed out. Figure 12.9 shows a typical cable scanning report.

The wire map tests are reported simply as Pass or Fail. The length measurement is performed on all pairs of the cable under test, but the pair with the shortest electrical delay is used for the Pass/Fail determination and its electrical length (in feet or meters) is reported. An additional 10% in length is permitted to allow for uncertainty in the NVP of the cable.

```
              LANcat V  CABLE CERTIFICATION REPORT #3

                        Company Name Here

  Circuit ID: H7/B6-61            Date Tested: 06/24/1996
  Cable Test Standard: CAT5 Basic Link   Module Type: Modular Plug
  Location: _____      Remote Module Type: TwoWay Remote
                                   Serial Number: 9510704  V2.41D1
                                   Cable NVP:  72.0%

  TEST SUMMARY: PASS
```

Wire Map: PASS	Near End	1	2	3	6	4	5	7	8	Shield
		|	|	|	|	|	|	|	|	open
	Remote End	1	2	3	6	4	5	7	8	

Length: PASS	Limit: 308ft		
	Pair	Length	Comment
	1,2	3ft	PASS
	3,6	3ft	PASS
	4,5	3ft	PASS
	7,8	3ft	PASS

Attenuation:	PASS	1,2	3,6	4,5	7,8	
		PASS	PASS	PASS	PASS	
Attenuation	dB	1.9	1.8	2.0	2.3	
Limit	dB	19.8	20.0	21.6	20.1	
Margin	dB	+17.9	+18.2	+19.6	+17.8	
Frequency	MHz	86.4	87.3	100.8	88.2	

Local NEXT:	PASS	12/36	12/45	12/78	36/45	36/78	45/78
		PASS	PASS	PASS	PASS	PASS	PASS
NEXT	dB	43.7	44.5	76.5	36.4	45.5	43.2
Limit	dB	31.0	29.3	60.0	29.4	32.2	29.3
Margin	dB	+12.7	+15.2	+16.5	+7.0	+13.3	+13.9
ACR	dB	41.9	42.5	76.5	34.4	43.8	41.9
Frequency	MHz	79.2	100.1	1.1	99.0	66.8	100.1

Remote NEXT:	PASS	12/36	12/45	12/78	36/45	36/78	45/78
		PASS	PASS	PASS	PASS	PASS	PASS
NEXT	dB	42.6	41.5	83.5	34.9	45.4	43.8
Limit	dB	29.5	30.9	60.0	29.3	32.9	29.3
Margin	dB	+13.1	+10.6	+23.5	+5.6	+12.5	+14.5
ACR	dB	41.5	39.9	83.5	32.9	43.9	42.5
Frequency	MHz	97.6	80.3	1.1	100.1	60.5	100.1

```
  Operator: _____    Date: _____

  Comments:
```

Figure 12.9 A typical cable scanning report. *(Courtesy of Datacom Technologies, Inc.)*

The tester must make a determination of attenuation Pass/Fail based on a calculation of the allowable attenuation at each frequency of measurement. If the cable fails, the attenuation at the frequency (or highest frequency) of failure is reported. Measurement steps are to be no greater than 1 MHz intervals up to 100 MHz for Category 5, or the maximum frequency for lower categories of cable. Attenuation values of less than 3 dB are reported but are not used in a Pass/Fail determination. An optional determination of the attenuation per unit of length is calculated for cables longer than 15 m and identified if they are greater than that expected. However, higher than expected attenuation per unit length values do not cause a failure to be reported, but might be a cause for concern.

The NEXT Loss measurement is made in steps of 150 kHz from 1 to 31.25 MHz, and 250 kHz from 31.25 to 100 MHz. For the NEXT test, the measurement of the worst pair combination is used. The tester may report either the actual NEXT or the NEXT margin, although NEXT margin is required for failure reports or when the measurement is within the tester's accuracy from the limits. NEXT margin is a value of much significance to cable installers because it indicates when marginal installation practices may be corrected by retermination or rerouting. Often NEXT can be reduced by more closely maintaining the twist of the pair as close to the point of termination as possible.

Test reports must indicate whether the cable under test has passed or failed each requirement. Because of the difficulty in establishing an absolute measurement near the performance limits for the cable link, a cable may be falsely accused of failure or improperly passed. The standard takes a clever way of dealing with this problem—essentially including the tester's accuracy as if it were a part of the cable under test and allowing an additional 10% of the horizontal length limit to allow for NVP uncertainty.

A Fail indication, thus, is a sure indication of a parameter that is outside the limit. A Fail* indicates that the tester measured a value that is outside the test limit but by an amount within the tester's accuracy. A Pass indication is assumed to be adequately below the limit, with the accuracy of the instrument taken into consideration.

However, a Pass indication that results from a test measurement that is within the accuracy level of the actual limit is marked with an asterisk (Pass*). The link is still considered to pass the test but is a handy visual clue to which links may be at or near the absolute limits. Prudent installers might want to take a second look at these links to see if shortening the length slightly or more carefully controlling untwist might produce a full Pass. Because the test-reporting procedures allow an additional length variation of 10% in measurement of the electrical length, to account for NVP variations, you should really respect the length limit of the tester.

Occasionally the source of a Fail indication will be the "short link" problem described earlier. Certain links under about 15 m may exhibit NEXT failure due to the resonance phenomenon. These links may actually be "cured" by replacing the run with a longer length of cable. The evaluation of such links is under study and will, no doubt, be the subject of a revision to TSB-67 or TIA 568-A at some future time.

Analog vs. Digital Field Testers

A great controversy exists between two measurement technologies for high-performance cable scanners. The two techniques are generally called "analog" and "digital." The analog testers use a conventional continuous waveform to test such link parameters as attenuation and near-end crosstalk (NEXT). The frequency of the waveform is changed (stepped or swept) and the measurement repeated at each of hundreds of discrete frequency steps that are required for full bandwidth testing. The digital testers measure these parameters by generating a series of pulses, then using digital signal processing techniques to derive the same parameters at all step frequencies more or less simultaneously.

The new testing requirements of *TIA / EIA TSB-67, Transmission Performance Specifications for Field Testing of Unshielded Twisted-Pair Cabling Systems*, push the performance limits of field tester technology. The TSB defines strict accuracy levels for the testers and defines two test configurations for testing a horizontal cabling link, the Basic Link and the Channel. In the bulletin, swept/stepped techniques are used to illustrate the measurements required of field testing for cable links. However, the TSB allows other methods "using frequency domain or time domain [digital/pulse/DSP] measurements" that are equivalent to the swept frequency [analog] methods.

In addition, TSB-67 follows the interesting practice to exclude the actual connection (for example, the plug and jack) which attaches the tester to the link under test. In the case of the Basic Link, this connector is not specified, as it is at the tester end of the 2 m (maximum) test cable. Some tester manufacturers use a high-performance, low NEXT connector at the tester so as to not introduce adverse levels of NEXT into the link under test. However, at the Channel interface, the actual user cable must connect to the tester at each end. The user cable must be terminated in the standard 8-pin modular plug, that unfortunately has relatively poor NEXT performance.

The most intriguing claim of the digital tester advocates is that the pulse/DSP technique allows them to totally exclude the contribution to NEXT and other parameters that results from the modular connector (mating plug and jack) at the tester itself. The modular connector is the required test interface for a Channel measurement, so an alternative low-NEXT connector cannot be used, as it could for the Basic Link. Remember that a very significant portion of the detrimental crosstalk occurs at this modular connector interface. Digital signal processing can actually exclude this connection measurement, unlike conventional analog methods.

Which type of field tester should you choose? Sorry, but we are going to leave this decision to the reader! The technical positions of each tester manufacturer are strong and emotions run high. There has basically been an agreement to disagree. You may wish to look at several different models and make your own judgment. Frankly, the many testers on the market actually differ in features, options, reporting format, price, and measurement speed in addition to the measurement technique. Either style of tester is acceptable, both to the TIA working committees and to qualified independent evaluators. The bottom line is, choose the specific tester that will do the best job for you.

13

Monitoring
and Administration

A LAN wiring system is a dynamic entity in many ways, changing and growing much like a living thing. If the LAN operating system is like the life blood of your network, the signal wiring certainly is like the nervous system, carrying important messages from point to point. LAN wiring must be monitored and tracked, so that you can constantly know the status of your interconnections.

LAN wiring systems are not truly passive systems. They are active in the sense that they can change characteristics over time and can disrupt the LAN if they fail. When a failure occurs, you will need to make quick repairs. How well you are able to respond to a cable problem depends upon the methods available to you to isolate the problem and find the offending cable system component. Proper documentation of your wiring will assist you in your installation and troubleshooting efforts. This is the essence of monitoring and administration.

Monitoring Methods

Monitoring of a cable system can be accomplished either directly or indirectly. For most installed LAN systems, the networking equipment (servers and workstations) can do a part of the job indirectly. We call this "indirect" monitoring, since it is a part of the normal operations of the connected systems, which sense the status of the network links indirectly. If a workstation cannot connect to a server, it may mean that there is a connectivity problem, such as a bad cable or other wiring component. Unfortunately, indirect monitoring of the channel may indicate a communications problem even when the cable connectivity is fine. For example, a bad workstation network interface or a software problem at either the workstation or server would appear as a network communication problem, even though the cabling was good.

"Direct" monitoring uses the capabilities of the hubs, patches, or network interface cards (NICs) to directly monitor the cable link. This type of monitoring senses the condition of a cable and allows direct control and/or testing of the channel. Direct monitoring may be either local or remote. Local monitoring simply uses the built in status indicators on hubs and NICs. To monitor link status, one simply looks at the LED indicators on the devices. Of course, this assumes that the devices you are using have indicator lights.

Remote monitoring uses intelligence built into the hubs or other network devices to report status to a monitoring station. Remote monitoring can use standard network monitoring software, such as SNMP, described later in this chapter. Direct monitoring is rather new and offers an exciting new view of the LAN wiring as a separable entity. It is, unfortunately, still somewhat too expensive to implement on the low-end networking devices. However, as network speeds increase, and as the costs of networking equipment decrease, it will become more sensible to use monitoring devices to determine the health of the wiring system.

Physical Cable Monitoring

Physical monitoring of installed LAN wiring presents some very special problems. Most of the simple ways that can be used require that the monitoring be performed by the connected LAN devices (such as Token-Ring or 10BaseT hubs). The hub monitors the condition of the link and reports it by indicator lights on the front panel of the hub, or by sending messages to a remote monitoring station. These devices and their capabilities are specific to the type of LAN you are using.

A Token-Ring MSAU, for example, uses dc connectivity on the link to set or reset a sensitive relay in the MSAU. Even a simple MSAU can easily provide an LED indicator on the front panel to indicate link condition. Intelligent MSAUs are able to report the status of any lobe to a remote monitor. Of course, the indication shows more than just the integrity of the cable, since the workstation operation is also a factor in lobe status. Nevertheless, this is an effective means of physical monitoring.

Likewise, the 10BaseT and the 100 Mbps Ethernet twisted pair systems provide a positive indication of link operation. Standard 10BaseT actually only evaluates the integrity of the received signal from the other end of the link, but enhanced systems can indicate integrity of the complete link and report to a remote monitor.

Some monitored patching systems can provide connectivity monitoring with virtually any type of LAN. These systems typically use pairs on the cable that are unused by the LAN. For example, both 10BaseT and Token-Ring use only two of the pairs on the cable. The other two are available for other purposes, in this case connectivity monitoring. While this does not truly indicate that the network pairs are connected, it certainly will indicate when a patch cable is disconnected. Some systems can even detect where a patch cord has been moved to and report an alarm to a monitoring station. The monitored patch system in Fig. 13.1 is an example of this capability.

Figure 13.1   The monitored patch system. *(Courtesy of RIT Technologies Ltd.)*

SNMP Monitoring and Control

Effective remote access to network devices may be provided using the Simple Network Management Protocol (SNMP). This protocol communicates over the network to a management workstation via an extension of the popular TCP/IP suite. Monitoring of all sorts of hubs, bridges, routers, and other devices can be accomplished using SNMP. SNMP management can be expanded to almost any device with a network connection. Some devices that have no native LAN ports, such as T1 communication equipment, are now being equipped with a LAN port just for an SNMP interface.

SNMP-based monitoring and control techniques may also be applied to cable management. Particularly with hub-based cable management, SNMP is a welcome tool for both management and control. Some of the intelligent patch panel equipment can communicate via SNMP for both monitoring and control. As networking hardware becomes more complex, global network management solutions will become more important. Hopefully, these management tools will offer some degree of cable management along with their other functions.

Remote Control of Physical Connections

In some circumstances, you may have need for remote control of your physical network connections. For example, you might want to disconnect or reconnect a workstation from your network monitoring location. This can be effective in preventing a single workstation from causing a global network problem.

Many intelligent hubs and similar devices can provide this remote control capability. The most common method is to use in-band signaling via SNMP or a proprietary protocol. However, this method does have one disadvantage. If your network is being jammed by an offending workstation, it may be impossible to communicate over the LAN to command the hub.

Another remote control option is an out-of-band control system using a control port. This control port remains free of any network disruptions and can keep you in the game. A control port can also be used to make a modem connection to a remote system, regardless of the network connectivity.

Some remote controlled equipment can perform other monitoring functions such as environmental monitoring and access control. A remote temperature alarm could alert you to a condition that might eventually cause the devices in an equipment room to fail. Or you could simply have a remote system start the coffee before you arrive.

LAN System Documentation

Good documentation is the key to controlling your LAN wiring environment. Whether you are troubleshooting a LAN outage or simply adding a cable drop, having well-organized records will save you a great deal of time and trouble.

The systematic marking and identification of all of the components of your LAN wiring system is an essential part of good documentation. Each item of the LAN wiring should be clearly marked with an identifying number. This applies to each cable, each termination, and all of the points in between. As a matter of fact, many of the mounting structures and pathways that are used in support of LAN wiring are best identified as well.

In an attempt to standardize the administration of telecommunications cabling, the EIA/TIA has published *EIA/TIA-606, 1993, The Administration Standard for the Telecommunications Infrastructure of Commercial Buildings*. This standard covers the identification, nomenclature, and documentation of wiring, pathways, and grounding/bonding in commercial buildings. It also suggests a system of drawings, work orders, and computerized record keeping to help manage the wiring system after initial installation. Figure 13.2 shows some of the wiring system elements that are expected to be identified and recorded. EIA/TIA 606 is quite detailed and specific, going far beyond the norm in some areas. However, it provides a very useful documentation and administration plan from which to model your record keeping.

The guiding rule of a proper LAN documentation system is "mark everything." That is, systematically identify every outlet, cable, termination block, patch panel, cable pathway, and telecommunications space. Then, record those identifications on drawings and in a database for quick reference.

If you do not carefully manage your LAN wiring installation, the constant pace of changes to the system will soon disorganize things so much that you

Figure 13.2 Many wiring system elements are expected to be identified and recorded.

will be forever lost in the cabling maze. The extent to which you need to implement the full complexity of the EIA/TIA 606 system depends upon the size of
your cable plant. Large systems need much more careful documentation than
small systems (under 50 stations). As a minimum, the wiring outlets and cable
terminations (patch panel jacks or punchdown blocks) need to be given identifying numbers. It is also a good practice to permanently label the ends of each
cable run.

Telecommunications outlets should be clearly numbered. The standard method is to place a label on the faceplate above each jack. Many manufacturers have responded to this requirement with special faceplates that have a recessed label area below a removable protective plastic cover. An example of these faceplates is shown in Fig. 13.3. Patch panel jacks should also be labeled on the front of the panel. Labeling on the back is optional, but is a very nice feature for troubleshooting.

Figure 13.3 An outlet faceplate with labeling panels. *(Courtesy of The Siemon Company.)*

TABLE 13.1 Items That Must Be Recorded in the Cable System Documentation

Record	Type	ID?	Type	Other info:	Linkage to records for:
Wiring spaces	Pathway	Y	Y	Fill, loading	Cable, space, pathway, grounding
	Space	Y	Y		Cable, pathway, grounding
Wiring	Cable	Y	Y	Unterminated, damaged, available pairs	Termination position, splice, pathway, grounding
	Termination	Y	Y	Hardware, Damaged position no.'s	Termination position, Space, grounding
	Position	Y	Y	User code, cable pair no.'s	Cable, term. position, term. hardware, space
	Splice	Y	Y		Cable, space
Grounding, bonding	TMGB	Y	Y	Grounding conductor, resistance to earth, date of measurement.	Bonding conductor, space
	Bonding cond.	Y	Y	Conductor	Grounding, pathway
	TGB	Y	Y		Bonding conductor, space

Source: EIA/TIA-606

Patch cords that run from cable terminations at the patch panel to the hub equipment ports are also numbered to make it easy to find which station cable corresponds to which hub port. Patch cords should be numbered with a removable label, if possible, as they are intended to be moved around. This practice makes it easy to identify the proper jack location for each cable end. Perhaps, in a very small installation, you will be able to maintain a one-to-one correspondence between the two patch jacks, but a larger facility will soon lose this neat arrangement. There is nothing worse than a loose patch cord end hanging down with no idea where to plug it back in.

Large installations will benefit from the labeling of cable pathways, such as ducts, sleeves, cable trays, and conduit. Raised floors, while not directly marked, can also be identified by drawings showing grid numbers for each floor area. Work area and telecommunications spaces should also be identified, with a marking at the entrance to the area. For many buildings, this requirement is easily handled by existing room numbering.

Additional information is recorded along with the identification number of each element in the wiring system. The exact information required varies with the type of item, but it always includes the identifier and the item type.

Additional information may be needed as to the numbers of available positions and whether any are damaged (as in bad pairs in a multipair cable). The administration standard also requires that the information recorded for each wiring element be cross-linked to any of the records that are related to that item. For example, a cable identifier must be linked to records of termination positions, splices, pathways, and grounding. A list of the item types, information to be recorded, and record links is shown in Table 13.1.

Automated database documentation systems can also be very useful. Some cable administration software has a graphical interface that lets you point to items in your cable plant and see the pertinent information that is recorded. A cable administration system can really be operated apart from the network management system (NMS) for your LAN. NMSs tend to concentrate on network node and hub management and control and are not concerned with what cable duct a particular pair might be located in. In addition, the cable administration software can cover telecommunications cable uses that are far beyond the LAN.

14

Troubleshooting

Any complex system is subject to periodic failures, and LAN wiring systems are no different. Throughout this book, we have emphasized the installation of a reliable wiring system. We have spent a great deal of time showing the proper components and installation techniques that are required to create a properly functioning, reliable cable plant. The precept for this careful construction is that when anything does go wrong, it should very rarely be the cabling. (Some might say "never" rather than "rarely," but your LAN wiring is in an environment where it cannot totally be protected from those uncaring souls without the proper respect for these thin little wires.)

The fact of the matter is that failures do happen. Fortunately, most of the failures happen during installation, where the link operation is not critical. Still, the failed cable must be repaired, regardless of the circumstances. If the link has been in operation, it is all the more important that we make the repairs quickly.

Cable link failures can be divided into two broad classes with respect to LAN wiring. The classes of failures correspond to the types of test equipment for each. Failures involving basic cable function can be determined by the basic test equipment we talked about in Chapter 12. This equipment can check cable link connectivity and dc wire map. Failures involving performance-based cable function must be determined by the diagnostic and measurement equipment we covered. This equipment tests the performance of the cable link at the level of performance required for your particular LAN application. It tests for split pairs, impedance, NEXT, length, attenuation, and other parameters that are required for proper LAN operation. The basic function can be fine, but the link will still not operate if it doesn't meet your performance needs. For example, you could probably get dc connectivity on a barbed-wire fence, but you wouldn't be able to run data on it.

In this chapter, we will detail many of the types of cable link failures, whether they are from installation errors or cable damage after installation. We will give some approaches to finding the problem and repairing the cable link. Also, we will say a little about a successful troubleshooting approach that you may find useful.

Common Failure Modes

Wire break

One of the most common wiring failures is a broken wire. Most testers will report the wire as an "open," meaning that the circuit is open. Some testers may actually declare the entire pair bad, although only one wire might be open. A wire that was never connected will also test open, even though the failure is not a result of wire damage. If only one wire of a multiwire cable is open, you should probably look for a bad connection at one of the termination points.

Connections in insulation-displacement connectors, such as 66M or 110 blocks, sometimes are not driven home, resulting in a bad or intermittent connection. What happens is that the insulation is not quite stripped away, and an apparently good connection will actually have a thin layer of insulating plastic remaining. You should never reuse old 66M blocks. The contact jaws can deform outward just enough to cause a problem when new wires are punched. This can also happen if a larger gauge wire was previously punched down. These blocks may also have the same problem with jumper wires that are frequently moved. You may be able to "rehabilitate" a contact by squeezing the empty contact jaws together with sharp-nosed pliers, or even by rocking the punchdown tool back and forth on the contact. The 110-type blocks are much less prone to this problem, as they use a cutting-displacement method of termination, where the connector contact actually slices though the wire's insulation.

If inspection of the cable termination ends does not reveal the open wire, use a cable scanner to measure the length to the open. Midcable open wires may be very hard to find visually, unless there is obvious mechanical damage to the cable. You never want to find out that an installed cable has an open due to a manufacturing defect, since it is almost impossible to prove that the break did not occur during installation. It is a good idea to do a continuity check on each reel of cable prior to installation. During installation, be careful not to pull the cables over sharp edges, as the cable might be cut. See the next section for advice on how to repair a twisted pair cable open or short.

Fiber optic cables may also be open or fractured, indicating that the cable has been severely bent at some point, or that a connector is bad. The only way to find the break, if it is not obvious by inspection, is to use an OTDR (Optical Time-domain Reflectometer). Fiber optic cables may be spliced by fusion of the two fiber ends at the break or by connectorizing both ends and using a coupler. Fusion of the fiber generally results in less loss, which is an important factor in an optical cable splice.

Wire short

A wire short in some ways is easier to deal with than an open. A short is rarely caused by a bad connection at a termination, so you can initially skip that inspection. A cable scanner may be able to locate the short within a few feet. If you do not have access to a scanner, use a sensitive ohmmeter and the specification for cable resistance per linear foot (26 ohms per 1000 ft for AWG 26 wire) to find the approximate location of the short. If more than two wires are shorted, this test may be useless, as the linear resistance will be much less than expected.

Shorts are often caused by other contractors trapping the cable between sharp objects during construction. For example, the cable might be pinched between metal studs as a wall is assembled. The ubiquitous sheet rock screw also does a very nice job of piercing right through a cable, shorting the wires in the process. Sometimes (very infrequently, we hope) cable installers will pull the cable past a sharp edge, cutting through the insulation and shorting wires in the cable.

The method used for repair of a short or an open depends upon the type of cable and the performance requirements. Coax cable may be cut in half, the two ends connectorized, and a barrel connector used to join the ends. Fiber may be spliced, as mentioned in the previous section.

Twisted pair cable repair methods depend upon the category of performance desired. Technically, the TIA 568-A standard allows no splices in a horizontal cable run. However, practical considerations may dictate some variance with Category 3 or 4 cable runs. Categories 3 and 4 have a more gentle twist than Category 5, and it is possible to carefully perform a splice using a shrink-wrap solder splice available from several sources. This hollow, shrinkable-plastic splice is about one inch long and contains a narrow ring of low-temperature solder in the middle. The insulation is removed on the wires to be joined. They are placed in the splice from opposite ends and the splice heated with a proper heat-gun tool to melt the solder and shrink the plastic. (A pocket lighter is not a proper tool, but it works well just the same if used with care.) The next conductor from the same pair is wrapped around the first wire in the same direction as the natural twist and the wire ends are joined in the same manner.

An alternative method will form a temporary splice, even for Category 5. TIA 568-A allows one transition point from round to flat cable in a horizontal cable run. This means that an allowance has been made for the inevitable untwist and NEXT that would result from terminating each cable end at the point of transition. Logically, you should be able to insert a plug and jack coupling to repair a cable cut or short, but that is not really allowed as it amounts to a splice. However, if a quick solution will get the network back running, this method can be used temporarily until new cable can be pulled. You should only use a properly rated plug and jack for the cable's category of operation. It may seem easier to use two jacks and an 8-pin coupler for this, but that is actually two plug/jack combinations, not one. Also, it is very unlikely that the 8-pin coupler will meet any standard higher than Category 3. In addition, many of these couplers also reverse the order of the pins, causing yet a different type of failure.

Finally, mark the point of the splice on your cable map. If the splice is temporary, you will want to go back and pull a new cable as soon as you can, as the need to replace the cable will be inversely proportional to how long you wait. If you want to have a Category 5 facility, but are only using Category 3 applications now, the splice may work now, but may fail when you upgrade to higher speed applications.

Kinks, bends, and breaks

The effect of kinks and bends in cable also varies with the type of cable. Coax cable is probably the most resistant to this type of damage, because of the size and durability of the cable. However, severe kinks and very sharp, sustained bends will affect the transmission loss on the cable and may even cause an impedance reflection that could cause a failure.

Fiber optic cable, on the other hand, may be permanently damaged simply by overflexing the cable. Fiber cable will typically show no outward sign of damage, but the optical fiber may have small cracks, called microfractures, that adversely affect link loss. Severe bending and sharp impact may even fracture the fiber completely. Breaks may also be caused by excessive pulling tension during installation. Damaged fiber cable will show poor loss performance even with the simplest fiber optic testers, but you will need an OTDR to find the location of the fracture. Fractures must be repaired using fiber splice techniques we covered in previous sections. Use particular care when pulling fiber not to kink, bend, or over pull the cable.

Twisted pair cable may also be subject to transmission impairment due to kinks or sharp bends. As unbelievable as it may sound, even tight tie-wraps may distort the cable enough to cause this problem. The real problem with kinks or bends in twisted pair cable is that the defect actually distorts the geometry of the cable. This can easily be observed on the screen of a TDR (Time-domain Reflectometer). Distorting the cable shows up as a reflection hump in the returned signal. Severe bending can actually cause a permanent signal impairment. How serious this impairment is will depend upon the category of operation and just how near the operating margins of the link you are operating.

A major source of kinks in cables is the use of boxed (rather than reeled) cable. These cable boxes are designed to allow the cable to feed in a spiral fashion, without the need for conventional cable spools and reel holders. Unfortunately, the cable can sometimes feed more than one loop of cable and the loops get mixed together trying to get out of the feed hole in the box. Other times, a single loop twists so that it does not feed in a spiral fashion, as the loop collapses, a kink results. Of course, kinks can occur easily in loose coils of cable that has been unboxed or unreeled. Try to avoid kinks, particularly Category 5 operation. If a kink occurs, attempt to smooth out the kink as much as possible, but be aware that you may have to replace the run if it later fails in testing.

The TIA 568-A standard recommends a minimum bend radius of four times the cable diameter. For most cables, that radius is about ¾ to 1 inch. Try to avoid pulling cable tightly around corners and use cable management retainers to provide a smooth transition to the point of termination.

Connector opens

Modular connectors may fail due to open connections. The common most conditions are bad plug crimps, dislodged wire, bent jack pins, and bad connector seating. The popularity of modular plugs in telephone wiring has resulted in the availability of low-cost, 6-pin crimp tools. These tools may be made from plastic or lightweight steel. Unfortunately, the flimsy design has been extended to 8-position crimpers. The 8-pin modular plug is much more difficult to properly crimp than plugs with 6 pins. The low-cost tools often cannot properly seat the connector contacts in the center of the plug. This may result in one or more of the stranded wires not making contact or, worse, making intermittent contact. You can easily identify plugs crimped with one of the inferior tools. When viewed from the front, the middle contacts are visibly higher than the contacts near the edge of the connector plug. Re-crimping rarely makes any difference, unless it is done with a proper crimp tool.

You should spend the extra money to purchase a high-quality crimp tool. These tools are heavily constructed and will be quite a bit more expensive than the low-cost tools. Some may have interchangeable dies for different sizes of modular connectors. A high quality crimp tool for 8-pin modular plugs should cost $100 to $200, depending upon additional features.

Bad crimps may also allow a wire to be pulled out of the plug contact area. The same type of problem can happen in a cable termination at a workstation outlet jack or at the patch or punchdown in the wiring closet. These connections are not made for very large pulling forces, and may fail when overstressed. User cords in the work area are particularly subject to this kind of damage, since they often are run along the floor in the path of chair legs, furniture, and feet. Always inspect both cable ends of a cord that is suspected of having a bad connection. Replace the cord if there appears to be damage.

A source of intermittent connections in modular plugs is the use of solid wire. These plugs were really designed for stranded wire and solid wire does not work well. Special versions of modular plugs are allegedly designed to work with solid wire, and they do work better, but you will be much better off if you only use stranded wire with these plugs.

In rare cases, modular connector jacks may not seat properly in the connector faceplate. Some designs include the plug channel in the molded plate, with a separate jack assembly that snaps into the plate. If the rear module is not seated properly, the jack wires may be too far back to properly connect with the plug contacts, even though the plug clip seats with a sharp click.

In rare cases, modular jack pins may get crossed or bent. This will usually test as a shorted wire from the far end and as an open connection from the near end. The design of these jacks places the pins (which are really bent wire pins) in narrow plastic slots so they can move as the plug is inserted. At the top of the tracks, most jacks have an open area where the pins could slide from side to side if the wire pin is pushed upward too far. If the pin is allowed to slip into another pin's track (either during assembly or use), the pins will short together and one or both of them may be prevented from making any contact at all with their corresponding plug contact. If you suspect this problem, visually inspect the plate and replace the jack if it is damaged. This defect tends to permanently bend the wire pin and you may have continuing problems. It is a good idea to glance at the pins in each jack insert before you terminate the station cables. This only takes a second, and will save you hours of frustration later.

Modular jacks should be mounted so that the pins are on the top side of the jack opening. The theory is that this keeps dust and dirt from contaminating the pins. Jacks are also available with dust covers, which should be left in place until the jack is used and replaced when a plug is removed.

A punchdown block or patch panel punch block is a special type of connector. Some older-style 66M punchdown blocks require the use of a metal clip to connect between contacts in different columns. The standard 66M block has four columns of contacts across each punchdown position. Generally the contacts are internally wired as pairs, so that Column 1 and Column 2 contacts are connected in common, as are Column 3 and 4. However, some connectorized designs make all four columns independent, so that a clip must be placed across the contacts in adjacent columns to complete the circuit. A missing clip will cause a problem with the circuit.

An additional problem with punch blocks of any type concerns the use of stranded wire. These side-displacement or insulation-cutting terminals are designed for solid wire only. Do not use stranded wire with these blocks. The wire may not make good contact, since the contacts press in from the side, rather than clamping the wire or piercing into the strands. Even if an initial contact is made, the wire may pull free more easily, since much of the strain resistance of the contact depends upon trapping a solid wire in the contact.

Occasionally, a connector block open can occur when the punchdown tool cuts the cable end of the wire being terminated, rather than the scrap end. A tool may sometimes cut an adjacent wire to the wire being terminated. It is a good idea to check the terminated wires to be sure none seems loose. Cut or loose wires should be reterminated.

Coax cables are subject to having their connections pulled open due to connector stress. Crimped or soldered center conductor pins are fairly sturdy, but the pin may sometimes be pulled back into the connector body and not make contact with the mating jack. A visual inspection will show a pin that seems too short. The center pin should be approximately even with the edge of the shield sleeve (not the bayonet sleeve). Cables that are pulled on while mated may also break the shield wires. If a coax shield is improperly stripped, many of the tiny wires that form the braid will be severed, with only a few remaining to make the connection. When pulled, these last few wires will break easily, leaving an open. The proper solution is to cut off the end, prepare the wire properly, and replace the connector.

Any coax LAN wiring that uses BNC-Ts or terminators may be subject to opens due to failure of the T or terminator. As a matter of fact, many technicians consider the T component to be the primary cause of coax network problems. The connector may fail by literally falling apart, or it may exhibit intermittent connections that just add to the frustration. Low cost connectors are the worst offenders. You would be wise to use only high-quality connectors, Ts, and terminators on these systems.

Wrong pinout

All of the wiring problems that we discussed in the test equipment in Chapter 12 are common failures of cable links. Among these are reversing the order of some of the wires, crossing two pairs, flipping the wiring order, splitting pairs, and total miswires. Reversing the order occurs in pairs. A common mistake is to reverse the primary and tracer colors of a single pair. Another common mistake is to mix up the green and brown pairs, or split these pairs, since the color difference may be subtle in some cable. If you try to count pin numbers in a modular jack, be sure to position the jack correctly (see Chapter 7 for a drawing). Flipping the wiring order of a user cord is a common error with flat (silver satin) telephone-style wire. Inspect the wiring order by holding both cord ends side by side. The colors (to the extent you can see them) should be in the same order straight across the plug. Better yet, use a cable checker.

The standard color code for 4 pair cables should be followed in wiring all jacks and wiring closet terminations. This code is different for punchdown blocks and modular jacks. Modular jacks actually have two color code standards, T568A and T568B. You must use the same standard throughout a horizontal cable run. This means that you must use the T568A wiring pattern at the workstation outlet if T568A was used at the patch panel termination. Keeping this straight is much more difficult if you use preassembled octopus (fan-out) cables for your equipment connection. Be certain which wiring pattern standard is used with the octopus cable and use the same at the workstation outlet jack. Using different wiring patterns results in a reversed-pairs indication on a tester. See Part II of this book for more details on these color codes and wiring patterns.

There are nearly as many ways to connect station wires to modular jacks as there are modular jack manufacturers. The most confusing jack wiring problems stem from the fact that the pairs are not connected to the pins of the modular jack in the same order as they are punched down in the wiring closet. For example, in the wiring closet, the cable is terminated on the 66M or 110 block in the pair order 1, 2, 3, and 4. That means that the blue-white Pair 1 gets punched down first on the first two pins of the cable position, the orange-white Pair 2 on the next two positions, and so on. At the modular jack, however, Pair 1 is terminated such that it connects to Pins 4 and 5. The orange-white or the green-white pair is terminated on Pins 1 and 2, depending upon whether you are using the T568A or T568B wiring pattern. If the terminations on the jack module are numbered the same as the pin order, you must be sure to place the wires into the correct slots before you terminate them. This order will be very different from that on the punchdown block. Fortunately, many jack modules have color-coded wire slots, so you can ignore the confusing position numbers.

If you choose to split your station cables into two jacks, with two pairs per jacks, you will have extra fun figuring out what color code to use at each jack. If you must do this, one approach is to call each pair in a 2 pair set Pair A or Pair B. Plan which pins on the modular jack should respond to each pair, and draw a connection diagram, skipping the unused wire positions on the jack. This avoids the obvious difficulty you would have placing your Pair 1 in the Pair 2/3 position at the jack. Of course, all of the color codes on the jack will be completely wrong, but you wanted to do it this way, didn't you?

A far better approach, and one that is allowed by TIA 568-A, is to wire up the jack with the normal T568A or B pattern and use an adapter cable external to the outlet jack. Simply make (or have made) an adapter that connects two 2 pair cables, terminated in a plug at one end, to a single modular plug, with each wire placed appropriately. That will split out the station cable to the two applications and avoid making your jack plate totally non-standard.

Wrong category

Another common failure of high performance applications is due to the use of the wrong category of cable and/or connectors. Either mistake will cause most performance-critical applications to fail. Remember that the category of performance is limited by the lowest category of component or cable used in a link. In addition, of course, you must maintain the distance below the limit, route the wire away from potential interference sources, and maintain the minimum twist of the pairs.

Many older cabling systems can support Category 3 operation. Cable that is 4 pairs and has a minimum of 2 to 3 twists per foot should operate, although not necessarily at the distance limit. Multipair cables of 25 or more pairs should be limited to 6 circuits (12 pairs) to minimize the effects of crosstalk that occurs on such cables. Analog telephone circuits can usually be accommodated on pairs in the same cable sheath, as their frequency of operation is much lower than LAN frequencies. However, you must use caution when placing digital telephone signals, including ISDN, in the same cable jacket as LAN data, as the two may very well interfere with each other. Many modern phone systems are digital systems, and fall into this area of caution. The best rule is to use only one application per cable, even with analog systems. A later upgrade of your phone equipment might inadvertently shut down your LAN.

Excessive untwist

The amount of untwisted cable that is allowable when operating at Category 3 is so great that it is easy to view the Category 5 requirements with skepticism. After all, how could you go from a generous allowance of two or more inches of loosely twisted wire to a ridiculous requirement of less than one half inch (13 mm)? Well, the answer is easily found when you consider that the maximum frequency of "interest" of the two categories jumps from 10 to 100 MHz. Obviously, the Category 5 requirements are much more stringent because they have to be to assure cable link performance at this level.

Good cabling practice is to limit the amount of untwist to only that required to terminate the wires. This is good practice whether you are terminating Category 5 or Category 3. Another good practice is to strip back only enough of the outer jacket as is needed to terminate the pairs. This maintains the cable twist and the positioning of the individual pairs in relation to each other. The better connectors and cable terminations have additional cable management devices to secure the wire in place and to minimize stress and bending of the wire.

Wrong impedance

Throughout this book we have talked about UTP cabling with 100 ohm impedance and STP cabling with 150 ohm impedance. Any mixing of these two types of wire in the same cable link will result in an impedance mismatch that may cause the link to fail. The different impedances cause a signal reflection at the point of intersection that will affect even lower-speed network links. At the higher speeds, the impedance match is critical to overall circuit performance. The standards generally allow for the expected variations in impedance from one lot of cable to the next. A variation of $+/- 15\%$ is allowed and more may be acceptable on short cable runs, although it would technically not meet the standard.

The real problem is that network equipment is designed to expect a cable impedance of 100 (or 150) ohms. Using a cable with a higher or lower impedance will cause less power to be coupled to the cable pair and result in a lower received signal arriving at the other end. If that end also sees an impedance mismatch, even more power will be lost in the transfer. Receiver performance depends upon a certain minimum signal strength to override the crosstalk interference from the near-end transmission. If a lower-than-expected signal strength is received, link performance will suffer and may even cause the link to fail.

One final note of caution: other twisted pair cables exist with different impedances than are in the standard. For example, in some parts of the world, 120 ohm twisted pair cable is widely available. This cable should not be mixed with 100 ohm cable in LAN wiring applications. Be sure your cable is the proper impedance for standard LAN installations.

Troubleshooting Approaches

Troubleshooting involves active problem solving in an operating environment. To those of us who do troubleshooting on a daily basis, the process may seem obvious. However, to one learning about complex system operation, or to one who has to supervise such operations, it may be useful to cover some of the important troubleshooting concepts.

If you had to reduce troubleshooting methodology to just four phrases, they would surely be as follows: observe the problem, logically divide the component parts, test each portion of the whole, and make no assumptions. The first three items are logical, sequential steps. The last item ("make no assumptions"), really should be included at every step. Let's examine each step one at a time.

First, you should observe the problem. Many times the problem you will be asked to troubleshoot will involve more than just cable and wiring. To distinguish the problem from normal operation, you should be familiar with the normal operation. If not, you should at least get a description of normal operation that is as detailed as possible. In many cases, the best source of this description is an equipment manual or an expert source. The user may be of little help. Here is an example. You are told by a user that the link to the server is down, but their computer is working fine in all other respects. You walk to the equipment room, find their link's hub connection (from your excellent documentation), and observe that the hub card status lights are all off. Familiar with the hub operation, you know that at least one light should always be on, even with the patch cable removed. You wouldn't really need to check the cabling, because you have found a hub card failure.

The next step is to logically divide the problem. Let us say that we found the hub card lights normal, except for the one indicating a bad link to our problem workstation. Now we divide the problem logically. A workstation computer connects to the wall outlet, horizontal cable runs to the wiring closet, the terminated cable is cross-connected to a patch panel, and a patch cord runs to the hub equipment. The hub and computer both look fine, so let's check the Channel. Disconnect the user cord (at the workstation!) and the patch cord (at the hub!), then test the cable for shorts or opens. The cable link was working before the failure, so we don't have to be too concerned about other miswires. Of course, if the application is a Category 5 100 Mbps one, we would want to use a cable scanner set for Category 5 tests, since cable operation at the higher frequencies could be impaired without affecting dc continuity. If the cable checks good, the physical connection is no longer suspect. Next, you could look at the computer setup, interface card, server setup, and all the other network items that could cause a logical connection to fail.

Another way to logically divide the problem is through substitution of known-good components. Alternatively, you could use substitution to verify the proper operation of suspected components including the computer and hub. For example, you could move a known-good computer to the suspected-bad outlet, or move the suspected computer to a known-good outlet. Substitution causes you to make a guess which component might be bad and then test your guess. It can be done with very little test equipment, but it requires you to proceed very methodically and with no preconceived assumptions.

The third step, testing each portion of the whole, you might notice has followed immediately after we logically divided the problem. You must be very certain what you test, or your tests may be invalid. For example, what if we had tested the Basic Link instead of the Channel in our prior example? Does that test all of the cable? No, it does not! If the problem had been a failed user or patch cord, we would have missed it.

So, be methodical, be sure you understand the problem, carefully structure your testing so that it eliminates as many variables as possible, and make no absolute assumptions.

A

Resources

Online Resources

The world of Local Area Networking and LAN wiring is constantly changing. The following pages contain information on many of the standards, standard-setting groups, and industry organizations that are concerned with LAN wiring. However, since the map is constantly being revised, the following online resources are available to assist you:

Internet/World Wide Web
http://www.lan-wiring.com

Newsgroups
comp.dcom.cabling
comp.dcom.lans.ethernet
comp.dcom.lans.token-ring

CompuServe Forums
Novell Hardware Forum
Windows Networking-A Forum
... and various manufacturer-specific forums

America OnLine (AOL)
Computing area, Telecom & Networking Forum:

Telecom message board categories
Networks/Lan & Networks/HW: Cabling Questions
Tips & Tricks/For Comm Systems Designers

Standards Organizations

U.S. Standards organizations

ANSI (American National Standards Institute)
11 W. 42nd St., 13th Floor
New York, NY 10036
212-642-4900
Originates, recognizes and promulgates standards. ANSI is the US member to the ISO.

ASTM (American Society for Testing & Materials)
100 Barr Harbor Drive
West Conshohocken, PA 19428-2959
610-832-9500
Develops standards and testing methods for materials.

ATM Forum
303 Vintage Park Drive
Foster City, CA 94404
415-578-6860 (info line)
415-688-4318 (fax on demand)
info@atmforum.com
Industry consortium for the development of ATM standards.

EIA (Electronic Industries Association)
2500 Wilson Blvd
Arlington, VA 22201
703-907-7500
Industry organization that provides standards for electronic components and equipment.

IEEE (Institute of Electrical and Electronic Engineers)
345 East 47th Street
New York, NY 10017
212-705-7900
Professional engineering association that develops standards for electronic technology. To order publications from IEEE Operations Center, see "Where to Get Copies of Standards" later in this Appendix.

NFPA (National Fire Protection Association)
1 Batterymarch Park
PO Box 9101
Quincy, MA 02269-9101
617-770-3000
Provides electrical code that is used for the basis of most local building codes. Generally recognized as the authority on permitted uses and proper practice in electrical wiring, including LAN cabling. To order publications, see "Where to Get Copies of Standards" in this Appendix.

TIA (Telecommunication Industries Association)
2500 Wilson Blvd.
Arlington, VA 22201
703-907-7700
An industry organization that provides standards for telecommunications components and equipment.

UL (Underwriters Laboratories, Inc.)
333 Pfingsten Road
Northbrook, IL 60022-2096
708-272-8800
Provides standards and testing services, including flammability standards and performance certifications for cable and connecting hardware.

International and Non-U.S. Standards Organizations

CSA (Canadian Standards Association)
178 Rexdale Blvd.
Rexdale (Toronto)
Ontario M9W 1R3, Canada
416-747-4000

ISO (International Standards Organization)
International Standards Organization
1, rue de Varembé
Case postale 56
CH-1211 Geneve 20, Switzerland
+41 22 749 01 11 (From U.S., dial international prefix 011 first)
+41 22 733 34 30 (fax)
central@isocs.iso.ch(e.mail)
A university-sponsored Internet site is available at
http://www.iso.ch/infoe/order.html

IEC (International Electrotechnical Commission)
International Standards Organization
3, rue de Varembé
Case Postale 131
CH-1211 Geneve 20, Switzerland
+41 22 919 02 11 (From U.S., dial internationsl prefix 011 first)

Selected LAN Wiring Standards

Standards organizations marked with an * do not supply copies of standards directly. Standards for these organizations may be obtained from the sources listed in "Where to Get Copies of Standards," later in this appendix.

- **ANSI (American National Standards Institute)**

    ANSI Z136.2. American Standard for the Safe Operation of Optical Fiber Communications Systems Utilizing Laser Diode and LED Sources.
    Note that ANSI also approves standards of other organizations, such as the EIA and TIA.

- **ASTM (American Society for Testing & Materials)**

    ASTM D 4565-90. Physical and Environmental Properties of Insulation and Jackets for Telecommunications Wire and Cable.
    ASTM D 4566-90. Electrical Performance Properties of Insulations and Jackets for Telecommunications Wire and Cable.

- **CSA (Canadian Standards Association)**

    C 22.2 0.3-1992. Canadian Electrical Code, Part 2
    CSA-PCC-FT4 and FT6. Premise Communications Cord, Physical Wire Tests
    CSA-T527. Bonding and Grounding for Telecommunications in Commercial Buildings (harmonized with EIA/TIA 607)
    CSA-T528. Telecommunications Administration Standards for Commercial Buildings (harmonized with EIA/TIA 606)
    CSA-T529. Design Guidelines for Telecommunications Wiring System in Commercial Buildings (harmonized with EIA/TIA 568)
    CSA-T530. Building Facilities, Design Guidelines for Telecommunications (harmonized with EIA/TIA 569)

- **EIA/TIA***

    Electronic Industries Association (EIA) and the Telecommunication Industries Association (TIA).
    The EIA and TIA are two separate organizations that provide standards for the electronic and telecommunications industries. The TIA was formed in 1988 from the EIA telecommunications group and the US Telephone Suppliers Assoc. The standards for telecommunications are jointly issued by both the EIA and TIA. The TIA now has responsibility for continued development of the former EIA/TIA standards. It is these standards that are important to LAN wiring. As these standards are revised, the organization names are reversed, thus EIA/TIA 568 has become TIA/EIA 568-A. Many of the EIA/TIA standards are coordinated or harmonized with other recognized national and international organizations, such as ISO, IEC and CSA. The telecommunications standards are also ANSI-recognized standards. All of these EIA/TIA standards shown are approved by the American National Standards Institute (ANSI).
    EIA/TIA-568 1991. Commercial Building Telecomunications Wiring Standard
    TIA/EIA-568-A 1995. Commercial Building Telecomunications Wiring Standard
    TSB-36 ** **1991.** Additional Cable Specifications for Unshielded Twisted-Pair Cable (incorporated into TIA/EIA-568-A)

TSB-40 [**] **1994.** Additional Transmission Specifications for Unshielded Twisted-Pair Connecting Hardware (incorporated into TIA/EIA-568-A)
TSB-53 [***]. Additional Cable Specifications for Shielded Twisted-Pair Cable
TSB-67 [**] **1995.** Transmission Performance Specifications for Field Testing of Unshielded Twisted-Pair Cabling Systems
EIA/TIA-569 1990. Commercial Building Standard for Telecomunications Pathways and Spaces
EIA/TIA-570 1991. Residential and Light Commercial Telecomunications Wiring Standard
EIA/TIA-606 1993. The Administration Standard for the Telecommunications Infrastructure of Commercial Buildings
EIA/TIA-607 1994. Commercial Building Grounding and Bonding Requirements for Telecommunications

- **IEEE (Institute of Electrical and Electronic Engineers).** *IEEE Standards 802.2, 802.3, 802.4, 802.5, and 802.6 are now ISO-approved and have been re-numbered to match the ISO numbering system.*

    IEEE/ISO 8802-3 (IEEE 802.3). Information Technology—Local and Metropolitan Area Networks—Part 3: Carrier Sense Multiple Access with Collision Detection (CSMA/CD) Access Method and Physical Layer Specifications (and Supplements J, K, L, and P-Q)
    IEEE/ISO 8802-5 (IEEE 802.5). Information Technology—Local and Metropolitan Area Networks—Part 5: Token Ring Access Method and Physical Layer Specifications (and Supplements B, C, and J)
    IEEE 802.12. Standard for Demand Priority Access Method. Physical Layer and Repeater Specifications for 100 Megabits per second Operation.

- **ISO/IEC* (International Standards Organization and International Electrotechnical Commission)**

    IEC 603-7. Connectors for frequencies below 3 MHz for use with printed boards.
    IEC 807-8. Rectangular Connectors for Frequencies Below 3 MHz Part 8: Detail Specification for Connectors, Four-Signal Contacts and Earthing Contacts for Cable, First Edition.
    ISO/IEC 11801 1995. Information technology—Generic cabling for customer premises.
    ISO/IEC 11801 corresponds to TIA 568-A. Additional ISO-approved standards are listed in the IEEE section. ISO standards may be ordered from the ISO member in each member country (ANSI in the U.S.) or from the document distributors listed in "Where to Get Copies of Standards", later in this Appendix.

[**] TSB, Formerly "Technical Systems Bulletin" (migrating to "Telecommunications Systems Bulletin" for TIA standard supplements).

[***] This TSB was not formally released, but was incorporated into TIA/EIA-568-A.

- NFPA (National Fire Protection Association).
  National Electrical Code, 1993, ISBN 0-87765-383-6
  National Electrical Code, 1996, ISBN 0-87765-402-6
  NEC Handbook, ISBN 0-87765-384-4
- Order from NFPA at 1-800-344-3555 or by ISBN from any bookstore.

- **UL (Underwriters Laboratories)**
  UL 910. Flamespread tests
  UL 1863. Standard for Communication Circuit Accessories

Membership Organizations

The following organizations are of general interest to those involved in LANs and LAN wiring:

BICSI (Building Industry Consulting Service International)
10500 University Center Dr., Suite 100
Tampa, FL 33612-6415
813-979-1991
800-242-7405
A membership organization for individuals and companies. Provides designer and installer certifications, building guidelines handbook.

EIA (Electronic Industries Association)
2500 Wilson Blvd.
Arlington, VA 22201
703-907-7500
An industry organization that develops standards for a wide range of devices, from semiconductors to complex electronic equipment. The original parent of the TIA.

IEEE (Institute of Electrical and Electronic Engineers
345 East 47th Street
New York, NY 10017
212-705-7900
A professional membership organization for individuals with extensive educational and standards activities. Develops standards for LAN networking of several types, including Ethernet, Token Ring, and AnyLAN.

NFPA (National Fire Protection Association)
1 Batterymarch Park
PO Box 9101
Quincy MA 02269-9101
617-770-3000
A membership organization for individuals and companies that develops fire safety and electrical standards. Publishes the National Electrical Code that is used as a basis for building codes worldwide.

TIA (Telecommunication Industries Association)
2500 Wilson Blvd.
Arlington, VA 22201
703-907-7700
An organization for the telecommunications industry. Develops for standards telecommunications wiring, including the TIA 568-A standard. Originally part of the EIA.

Where to Get Copies of Standards

Direct sales from standards organizations:

- **ANSI and ISO standards**

  American National Standards Institute
  11 W. 42nd St., 13th Floor
  New York, NY 10036
  212-642-4900

- **ATM Forum standards**

  The ATM Forum
  303 Vintage Park Drive
  Foster City, CA 94404
  415-578-6860 (Info line)
  415-688-4318 (Fax on demand)
  info@atmforum.com

- **IEEE standards**

  IEEE Operations Center
  445 Hoes Lane
  PO Box 1331
  Piscataway, NJ 08855-1331
  908-562-0060
  800-678-4333

- **NFPA standards**

  NFPA Headquarters
  1 Batterymarch Park
  PO Box 9101
  Quincy MA 02269-9101
  617-770-3000
  800-344-3555

- **UL standards**

  Underwriters Laboratories, Inc.
  333 Pfingsten Road
  Northbrook, IL 60022-2096
  708-272-8800
  800-676-9473
  800-786-9473

- **CSA standards**

  Canadian Standards Association
  178 Rexdale Blvd., Rexdale (Toronto)
  Ontario M9W 1R3, Canada
  416-747-4044 -orders
  416-747-4000 -main

Most of the standards from these organizations and from EIA, TIA, ANSI, ISO, ASTM, CSA and other organizations are available from the following document distributors:

Document Center
1504 Industrial Way, Unit 9
Belmont, CA 94002
415-591-7600
http://www.doccenter.com/doccenter/

Global Engineering Documents
15 Inverness Way East
Englewood, CO 80112-5766
303-792-2181
800-854-7179 (US/CAN)

For additional information on availability of standards documents, see the standards sections at **http://www.lan-wiring.com.**

Glossary

10Base2 The IEEE/ISO 8802-3 (IEEE 802.3) standard for Thin Ethernet coax networks—10 Mbps transmission, baseband signaling, 185 meters per coax segment. Also known as Thinnet or Cheapernet.

10Base5 The IEEE/ISO 8802-3 (IEEE 802.3) standard for Ethernet backbone (Trunk) Cable networks—10 Mbps transmission, baseband signaling, 500 meters per coax segment. Also known as Thicknet.

10BaseFL, FB, FP The IEEE/ISO 8802-3 (IEEE 802.3) standards for fiber optic Ethernet network connections for interrepeater links (See FOIRL), synchronous links, and passive links—10 Mbps transmission.

10BaseT The IEEE/ISO 8802-3 (IEEE 802.3) standard for unshielded twisted pair Ethernet networks—10 Mbps transmission, baseband signaling, unshielded twisted pair cable. Maximum allowable cable length is 100 meters.

100BaseFX The IEEE/ISO 8802-3 (IEEE 802.3) standard for fiber optic Ethernet network connections—100 Mbps transmission.

100BaseTX or T4 The IEEE/ISO 8802-3 (IEEE 802.3) standard for unshielded twisted pair Ethernet networks - 100 Mbps transmission, baseband signaling, unshielded twisted pair cable. Maximum allowable cable length is 100 meters.

100VG-AnyLAN The IEEE 802.12 standard for a token-control, arbitrated network connections—100 Mbps transmission. Media may be UTP or fiber optic.

ac Abbreviation for alternating current.

Alternating Current An electric current that cyclically reverses the direction of electron flow. The frequency is the rate at which a full cycle occurs in one second. (See current and direct current.)

ACR Attenuation Crosstalk Ratio. The value of attenuation, less the crosstalk value, both expressed in dB, at a particular frequency. A quality factor for cabling that expresses the relation of two important measured values. (See also S/N.)

Adapter, Fiber Optic A passive coupling device which connects two male fiberoptic connectors, aligning the light path.

Aerial Cable Telecommunications cable intended for installation between supporting points, such as poles or building entrance supports. Aerial cable normally uses a metallic strength member which may be internal, bonded externally to the cable jacket, wrapped along the cable, or otherwise attached to the cable to be supported.

Ampere A unit of measure of electrical current. The amount of current that flows through a one-ohm resistance when a 1-V electromotive force is applied. An ampere of current is produced by 1 coulomb of electrical charge passing a point in 1 second.

Amplitude The maximum value of a signal.

Analog Signal An electrical signal which varies continuously, without discrete steps, as with a digital signal. For example, the conversion of voice to a continuously varying voltage by a microphone produces an analog signal. (See digital signal.)

Anomaly An impedance discontinuity on a transmission cable. The discontinuity causes an undesired signal reflection which may distort signals and disrupt a network connection. Anomalies usually are often by distortions in cable geometry which do not affect the dc continuity of the cable.

ANSI American National Standards Institute.

Appletalk A network communications protocol developed by Apple Computer Corporation. Appletalk can operate over a LocalTalk cabling system at 230.4 kbps or over Ethernet at 10 Mbps.

Approved Ground A building ground or bonded connection to building ground that meets the requirements specified in the NEC.

ARCNET Attached Resource Computer Network. A token bus local area network standard developed by Datapoint Corporation. An RF signal is used to pass a token between stations connected in a bus or star. Star arrangements may use passive or active hubs for signal distribution. ARCNET can use a variety of cabling types, including coax, twisted pair, and fiber optic cable. Standard ARCNET operates at 2.5 Mbps.

ARCNETPLUS A version of ARCNET that operates at 20 Mbps. (See ARCNET.)

ATM Asynchronous Transfer Mode. A transport protocol based on fast switching of 53-byte cells. With its high-speed operation, fast switching, and guaranteed delay times, ATM can transmit voice, data, and video efficiently along the same transmission path. Conventional data access methods, such as Ethernet and Token-Ring, are supported by LAN Emulation (LANE), in order to compensate for their longer packet lengths. ATM is implemented on a variety of transmission media and at a variety of speeds. It may be used for both backbone and primary workstation connections.

Attenuation The decrease in magnitude of a signal as it travels through any transmitting medium, such as a cable or circuitry. Attenuation is measured as a ratio or as the logarithm of a ratio (decibel).

Audio Frequency The range of frequencies within the range of human hearing. This range is generally given as 20 to 20,000 Hz, although few humans can detect sounds below 30 Hz or above 16 kHz.

AWG American Wire Gage. A wire diameter specification. The smaller the AWG number, the larger the wire diameter.

Backboard   A wooden board attached to a wall for the purpose of mounting cable terminations and telecommunications hardware. Typically ¾ in plywood from 2 ft × 4 ft to 4 ft × 8 ft size. The backboard may be painted to match room decor or to indicate the function of wiring components that are mounted.

Backbone Cable   Cable which interconnects wire distribution centers in a building or a group of buildings.

Balanced Line   A circuit consisting of cable with two conductors which carry voltages of opposite polarity and equal amplitudes with respect to ground. The conductors must be twisted together to maintain balance over a distance. (See also unbalanced line, twisted pare.)

Balun   A transforming device that matches an unbalanced circuit to a balanced circuit. The balun may also transform impedance to maximize power transfer and minimize reflections.

Bandwidth   The range of frequencies required for proper transmission of a signal, expressed as a difference in hertz. A continuous range from zero is said to be a baseband signal, while a range which starts substantially above zero is said to be broadband (or narrowband in the case of RF signals).

Bel   A unit that represents the logarithm of the ratio of two power levels. (See dB.)

Bend Loss   Increased attenuation in an optical fiber caused by excessively small bend radius. Bend loss may be temporary or permanent if the microfractures caused by the bend attenuate the light signal through the affected area.

Bend Radius   The radius of curvature of a bend in a cable. Cables may have transmission performance degradation from a bend radius that is below a certain amount.

BICSI   Building Industry Consulting Service International.

Bifurcated   Split in two.

Binder   A tape or thread used for holding groups of pairs together within a cable. Multipair cables often bind the pairs in groups of 25 and 100 pairs.

Bit   One binary digit.

BNC   A coaxial cable connector that uses a "bayonet" style turn-and-lock mechanical mating method. Intended for RG-58 and smaller diameter cable.

Bonding   The method used to produce good electrical contact between metallic parts. In the wiring context, bonding describes electrical connection of grounding bars and straps within a building to the central approved ground.

Braid   Fine wires woven into a cylindrical cable layer which shields enclosed conductors. Braid also may be formed into a flattened conductor to be used as a grounding strap.

Buffer   A protective coating over a strand of optical fiber.

Butt-in Set   A telephone testing device resembling an oversized handset that is used for temporary connection and operation of a telephone line.

Byte   A group of 8 bits.

c   The symbol for the speed of light in a vacuum.

Cable   A group of insulated conductors enclosed within a common jacket.

Cabling   Installed cables. Also, the process of connecting cable between various points in a building or outside plant.

Capacitance   The ability to store electrical charge between two conductors separated by a dielectric material. In one type of capacitor, foil conductor sheets are separated by a thin insulating plastic layer and the assembly is tightly rolled into a cylinder, which is then encapsulated in a small metal can. In wiring, the capacitance is said to be "distributed," as the effect exists between wire conductors insulated with plastic that acts as the dielectric. Capacitance is specified in fractional farads (such as microfarads or picofarads).

CDDI   Copper Distributed Data Interface. A version of the FDDI access method that uses copper wire media, rather than fiberoptics.

Characteristic Impedance   A value of impedance (resistance and reactance) of a transmission line measured over a frequency range that would exist if the line were infinite in length. A transmission line of finite length will have perfect power transmission, allowing for absorptive transmission losses, if driven and terminated by an exact conjugate matching load impedance. An inexact match will cause reflections that increase transmission loss.

Cheapernet   Another name for Thinnet.

Chromatic Dispersion   (See dispersion.)

Circuit   A system of conductors which passes a signal or voltage between two points.

Coax   (See coaxial cable.)

Coaxial Cable   A type of cable consisting of a conductor centered inside a tubular conducting shield, separated by dielectric material and covered with an insulating jacket. The shield may be a foil, wire braid, or solid metal, and is typically at ground potential.

Collision   A simultaneous transmission by two stations on a shared medium. In practical terms, a condition where a second device begins to transmit before another device's transmission has completed.

Conductivity   The ability of a material to allow the flow of electrical current. Conductivity is the reciprocal of resistivity. The unit of measurement is the mho (ohm backwards).

Conductor   A material that offers low resistance to the flow of electrical current.

Cord   A very flexible insulated cable. User and equipment cords are typically made with stranded conductors and are terminated with the proper connector plug.

Core   The light transmitting central portion of an optical fiber. The core is surrounded with a cladding with has a higher refractive index and thus helps channel the light along the core, as a waveguide.

Coupling   The transfer of energy between two components of a circuit or cable. The unwanted transfer of signals between pairs of a cable or between a cable and outside objects.

CPE   Customer Premise Equipment. Equipment that is usually provided by the user or subscriber of a telecommunications service. (See customer premises.)

Crossed Pair   A wiring error in twisted pair cabling where both conductors of one pair are incorrectly exchanged for conductors of another pair at one end of a cable.

Crosstalk The coupling of unwanted signals from one pair within a cable to another pair. (See NEXT.)

CSA Canadian Standard Association.

CSMA/CD Carrier Sense Multiple Access with Collision Detection. The access method used by Ethernet.

Current The flow of charge in a conductor. Current flow is usually considered the flow of positive charge, whereas electron flow is the flow of negative charge. (See alternating current and direct current.)

Customer Premises Buildings, offices, and other structures under the control of the telecommunications customer.

dB DeciBel abbreviation. A logarithmic unit of measure expressing the ratio between two power levels. The value is 10 times the value in Bels. (See bel.)

dc Direct current. (See direct current.)

DEC Digital Equipment Corporation.

Demarc Demarcation Point. The point of telecommunications circuit termination by a telecommunications carrier within a building or residence. The user (subscriber or customer) has the responsibility for circuit operation beyond the demarc.

Dielectric An insulating material between two conductors.

Dielectric Constant The ratio of the capacitance of a capacitor whose plates are separated by the dielectric material to a capacitor whose plates are separated by a vacuum dielectric. Permittivity.

Digital Signal An electrical signal that uses two or more discrete physical levels or signal phases to transmit information. For example, the conversion of a voice waveform to a digitally encoded representation of the waveform produces a data stream in which instantaneous amplitude samples of the voice are represented by a binary value. The data stream may then be transmitted as discrete elements and decoded at the far end to produce the original analog signal.

Direct Current An electric current that flows in one direction and does not reverse the direction of charge flow. (See current and alternating current.)

Dispersion The phenomenon in an optical fiber whereby light photons arrive at a distant point in different phase than they entered the fiber. Mode dispersion occurs when elements of the optical signal take slightly different paths along the fiber and the differing path lengths cause the received signal to lose definition. Chromatic dispersion is caused by differing transmission times of different wavelengths of light, which are refracted differently, according to each wavelength. Dispersion causes receive signal distortion that ultimately limits the bandwidth and usable length of the fiber link.

DIX Abbreviation for DEC/Inter/Xerox, the original consortium that developed the standard for classic Ethernet 1.0.

Drain Wire An uninsulated wire in contact with a shield braid or foil along its length. At terminating points of the cable, the drain wire may be used to connect to the shield.

E Electronic symbol for voltage, the electromotive force.

E1   A specially framed 2.048 MBps wide area transport system which uses two twisted pairs for transport between a telephone central office and a customer premise. Similar to T1, but with 32 channels at 64 kBps. The primary deployment in European telecommunications systems. The system was originally designed for digitized voice transmission, but is often used to link digital data systems, such as LANs.

Earth   A term for zero-reference ground.

EC   European Community.

EIA   Electronic Industry Association. A membership association of manufacturers and users that establishes standards and publishes test methods. (See TIA.)

Equipment Cord   (See patchcord.)

Electromagnetic Field   The combined electric and magnetic fields caused by electron motion in conductors. EM fields may exist at great distances from the conductors, however-near-field effects are of more concern in wiring.

Electrostatic   The electrical charge which exists when the charge is at rest.

Electromagnetic Interference   An interfering electromagnetic signal. Network wiring and equipment can be susceptible to EMI as well as emit EMI. (See radio frequency interference.)

Equipment Room   ER. An area or that contains telecommunications or LAN equipment. An ER may be collocated within a telecommunications (wiring) closet.

EMI   (See Electromagnetic Interference.)

Ethernet   A LAN topology and access method that uses the CSMA/CD transmission method. Ethernet may be used with two types of coax cable, twisted pair, or fiber optic cable. Two varieties of Ethernet exist, see text.

f   Frequency.

FCC   Federal Communications Commission.

FDDI   Fiber Distributed Data Interface. A LAN topology and access method that passes tokens over dual counter-rotating fiberoptic rings at 100 Mbps.

FEP   Fluorinated ethylene-propylene. A thermoplastic with excellent dielectric properties which is often used for conductor insulation in fire-rated cables.

Fiber   (See optical fiber.)

Fiber Optics   The transmission of light through optical fibers for communications purposes.

Firestop   A material that prevents the passage of flame or smoke through an opening in a wall or floor.

FOIRL   Fiber Optic Inter-Repeater Link. An Ethernet fiber optic connection method intended for connection of repeaters. FOIRL is standardized as 10BaseFL and is defined in IEEE/ISO 8802-3 (IEEE 802.3).

Frequency   The number of times a periodic action occurs in a unit of time. The unit of measure is the hertz, abbreviated Hz. One hertz equals one cycle per second.

Frequency Response A range of frequencies within which a device operates within expectations. The limits are usually given as the frequencies for which the normal signal transfer of a device falls 3 dB below the nominal level. For example, an audio amplifier may have a frequency response of 20–20,000 Hz, while a telecommunications circuit may have a frequency response of 300–3000 Hz.

Giga Prefix meaning one billion (one thousand millions) (10 raised to the 9th power).

Gigahertz (GHz) A unit of frequency equal to one billion hertz.

GIPOF Graded-Index Plastic Optical Fiber.

Ground An electrical connection a point of zero potential, relative to earth ground. Earth ground is generally obtained through one or more grounding rods driven into the earth. Often called safety ground. Signal ground is a common return to a point of zero signal potential and may differ from the safety ground at the metal chassis in equipment.

Ground Loop A condition whereby an unintended connection to ground is made through an interfering electrical conductor.

Hertz The unit of frequency, one cycle per second. (See frequency.)

Home Run A cable run that connects a user outlet directly with the telecommunications closet, with no intermediate splices, bridges, taps, or connections.

Horizontal Wiring Telecommunications cabling from the user station to the first termination point in the telecommunications closet.

HVAC Heating, ventilation, and air conditioning system.

Hz Abbreviation for hertz.

ICEA Insulated Cable Engineers Association.

IDC Insulation-displacement connector.

Impedance The total resistance and reactance offered by a component. The unit of impedance is the ohm and is specified with resistive and reactive values as a complex number. The common symbol for impedance is Z.

Impedance Match A condition where the impedance of a one device is a conjugate (equal resistive component and opposite reactive component) match to the device to which it is connected.

Innerduct A flexible plastic raceway for protection of optical fiber cables. Innerduct is usually round and corrugated.

Insulator A device or material which has an extremely high resistance to current flow.

Insulation A material which is nonconductive to the flow of electrical current. The coating (usually thermoplastic) of a conductor which insulates it from other conductors.

Interference Undesirable signals which interfere with the normal operation of electronic equipment or an electronic transmission.

ISDN Integrated Services Digital Network. A transmission system for digital circuit distribution. The Basic Rate Interface (BRI) uses one pair for subscriber delivery of two 64 kbps "bearer" (user data) and one 16 kbps "data" (used for signaling and sometimes user data) channels. The Primary Rate Interface (PRI) delivers twenty-three 64 kbps bearer channel and one 64 kbps data channel.

ISO Abbreviation for International Organization for Standardization, the actual name in English, although the ISO abbreviation is often translated as International Standards Organization.

Jacket The outer protective covering of a cable.

Jumpers Unjacketed cables of one or more twisted pairs, used to crossconnect circuits terminated at punchdown blocks; also, connectorized lengths of 25-pair jacketed cable.

Kilo Prefix meaning thousand (10 raised to the 3rd power).

LAN Local Area Network. A computer network that typically exists within a single building or office or among a group of nearby buildings.

Laser A coherent source of light with a narrow beam. The word comes from "Light Amplification by Stimulated Emission of Radiation." Laser sources are used for single-mode fiber optic transmission. The laser source for most LAN connections is a semiconductor laser diode.

Lay The distance required for one conductor or conductor strand to complete one revolution about the axis around which it is cabled. Sometimes referred to as the "lay length" or "twist pitch."

Leakage An undesirable passage of current over the surface of or through an insulator.

LEC Local exchange carrier (the local telephone company).

LED Light-emitting diode.

Line Voltage The value of the electrical potential power line, generally given in volts-RMS.

Link Pulse A single-bit pulse which is transmitted every 2–150 milliseconds during idle periods on a 10BaseT link segment to verify link integrity.

Load A device that consumes power from a source.

Lobe An arm of a Token-Ring which extends from a MSAU to a workstation adapter.

Loss The portion of power from a source that is dissipated before being consumed by the load.

Manchester Coding A method of LAN signal coding that maintains a null dc level so as to allow the detection of collisions; the coding method of Ethernet and certain other networks.

MAU Media Attachment Unit. The transceiver in an Ethernet network; also, a common name for the MSAU in a Token-Ring network.

Mbps Megabits Per Second. A bit is one unit of a digital signal.

Media Filter An impedance matching arrangement used in Token-Ring networks to transform between the 100 ohm impedance of UTP and the 150 ohm impedance of media interface connections.

Mega Prefix meaning million (10 raised to the 6th power).

Megahertz (MHz) Unit of frequency equal to one million hertz (one million cycles per second).

Micro Prefix meaning one-millionth (10 raised to the –6th power).

Microfared One-millionth of a farad (commonly abbreviated μf, and less commonly μfd, mf, and mfd).

Micron (μm) Millionth of a meter (10^{-6} meter).

Mode A single electromagnetic wave travelling in a waveguide, such as an optical fiber.

Mode Dispersion (See dispersion.)

MSAU Multi-station Attachment Unit. The device for interconnecting station lobe cables to form a ring in a Token Ring network.

Multimode Fiber A fiber wave guide which supports the propagation of multiple modes. Multimode fiber may have a typical core diameter of 50 to 100μm with a refractive index that is graded or stepped. It allows the use of inexpensive LED light sources and connector alignment and coupling is less critical than single mode fiber. Distances of transmission and transmission bandwidth are less than single mode fiber due to dispersion of the light signal. (See also Single Mode Fiber.)

N Connector (See type N connector.)

Nano One thousandth of one millionth (10 raised to the–9th power).

Nanometer (nm) One billionth of a meter.

Nanosecond One billionth of a second.

NEC National Electrical Code.

NEMA National Electrical Manufacturers Association.

Neoprene A synthetic rubber which has good resistance to chemicals and flame. The chemical name is polychloroprene.

NEXT Near-end crosstalk. Crosstalk between two twisted pairs measured at the same end of the cable as the disturbing signal source.

NFPA National Fire Protection Association.

Noise An extraneous, random signal which interferes with the desired signal. In most measurement situations, a nonrandom undesired signal is counted as noise.

NVP (See nominal velocity of propagation.)

Nominal Velocity of Propagation The speed of signal propagation through a cable, expressed as a decimal fraction of the speed of light in a vacuum. Considered to be "nominal" because variations in geometry along a cable may produce absolute variations in the value. The NVP is used in reflectometry to closely estimate the length of a section of cable using the formula, Length = $\frac{1}{2}t \times \text{NVP} \times c$, where t is the transit time for the reflected pulse and c is the speed of light in a vacuum. NVP is sometimes given as a percentage, but must be converted to a decimal fraction to be used in the formula.

Ohm The electrical unit of resistance. The value of resistance through which a potential difference of one volt will cause a current flow of one ampere.

Open A break in the continuity of a circuit.

Optical Fiber A thin filament of glass used for the transmission of information-bearing light signals.

Outside Plant   Cabling, equipment, or structures that are out of doors.

OTDR   A version of a TDR intended for use with fiberoptic cables. (See TDR.)

Packet   A group of bits grouped serially in a defined format, containing a command or data message sent over a network.

Patch Panel   A panel containing jacks or connectors that is used to connect or provide access to telecommunications or data circuits.

Patch Cord   A flexible piece of cable terminated at both ends with plugs. Also referred to as "user cord," "equipment cord," or simply "cord" in conjunction with LAN use.

PCC   Premises Communication Cable, CSA (Canadian Standards Association) Cable Designation.

Pico   Prefix meaning one-millionth of one-millionth (10 raised to the −12th power).

Plenum   Any space, whether closed or open, that is used for air circulation. For example: air ducts, return ducts, air shafts, above-ceiling air return spaces. Work spaces are generally not considered plenums.

Plenum Cable   Cable which has been certified by the manufacturer as meeting accepted standards for installation in plenum spaces without an enclosing metallic conduit. Plenum cable is generally fire resistant and has low emission of smoke and toxic fumes in contact with a flame.

Polymer   A substance made of repeating chemical units or molecules. The term may refer to general types of plastic or rubber.

Polypropylene   A thermoplastic material similar to polyethylene but somewhat stiffer and with a higher softening point.

Polyethylene   A thermoplastic material with excellent electrical properties.

Polyurethane   Broad class of thermoplastic polymers with good abrasion and solvent resistance. Can be solid or cellular (foam).

Polyvinyl Chloride (PVC)   A general purpose thermoplastic used for wire and cable insulation and jackets. PVC is known for high flexibility. Often used in nonplenum wire insulation and cable jackets. A modified version of the material may be found in jacketing of some plenum-rated cables.

POTS   Plain Old Telephone Service.

Prewiring   Wiring installed before walls and ceilings are enclosed.

Protector   A device that limits damaging voltages on metallic conductors.

PVC   polyvinyl chloride.

R   Symbol for resistor. The Greek symbol Omega (Ω) is used to indicate a resistance value in ohms.

RF   Abbreviation for radio frequency. (See radio frequency.)

Raceway   A channel used to enclose telecommunications and electrical wiring. Raceways may be metallic or nonmetallic and may totally or partially enclose the wiring.

Radio Frequency   The frequencies in the electromagnetic spectrum that are used for radio communications. Generally considered to be frequencies above 300 kHz.

Radio Frequency Interference Electromagnetic interference at radio frequencies. (See electromagnetic interference.)

RCDD Registered Communication Distribution Designer. A certification of BICSI, an industry organization, for individuals qualified to consult and design in telecommunications distribution systems. (See BICSI.)

Receiver A device whose purpose is to capture transmitted signal energy and convert that energy for useful functions.

Reflection A return of electromagnetic energy that occurs at an impedance mismatch in a transmission line, such as a LAN cable. Reflections in an operating circuit are undesirable and may result from a physical problem, such as crushing or untwist, or from an electrical connectivity problem, such as a short or an open.

Refractive Index The ratio of the speed of light in a vacuum to its velocity in a transmitting medium, such as an optical fiber core.

Repeater A device that connects two segments in an Ethernet network, or two portions of a Token-Ring network ring so as to extend and regenerate the LAN signal.

Resistance The opposition a conductor offers to current flow in a dc circuit. Resistance is measured in ohms, where one ohm is the resistance which allows one ampere to flow when a potential of one volt is applied. Resistance is the real component of impedance, which includes opposition to current flow resulting from capacitive and inductive reactance.

Reversed Pair A wiring error in twisted pair cabling where the conductors of a pair are reversed between connector pins at each end of the cable.

RFI Abbreviation for Radio Frequency Interference. (See radio frequency interference.)

RG/U RG is the common military designation for coaxial cable and U stands for universal.

Ring A polarity designation of one wire of a pair indicating that the wire is that of the secondary color of a 5-pair group (e.g., the blue-white wire of the blue pair); a wiring contact to which the ring wire is connected; the negative wiring polarity. (See also tip.)

Riser Cable A type of cable used in vertical building shafts, such as telecommunication and utility shafts or between utility rooms via cable ports. Riser cable typically has more mechanical strength than general use cable and has an intermediate fire protection rating.

RJ-45 A USOC code specifying a particular analog line connection that uses an 8-pin modular telecommunications connector jack. The term RJ-45 is often used simply to identify the 8-pin, 8-wire/contact modular plug or jack that is used with twisted pair cable.

Rubber (Wire Insulation) A general term used to describe wire insulations made of thermosetting elastomers, such as natural or synthetic rubbers, neoprene, butyl rubber, and others.

Runt Packet An Ethernet data packet which is shorter than the valid minimum packet length, usually caused by a collision.

Scanner A cable testing device which uses TDR methods to detect cable transmission anomalies and error conditions.

Screened Cable   (See shield.)

Segment   A portion of an Ethernet network containing workstations, but no workstations, other than at the endpoints; a network cable terminated at both ends.

Sheath   (See jacket.)

Shield   A metallic foil or multiwire screen mesh that is used to prevent electromagnetic fields from penetrating or exiting a transmission cable. Also referred to as a "screen."

Shield Coverage   The physical area of a cable that is actually covered by shielding material, often expressed as a percentage.

Short   A near-zero resistance connection between two wires of a circuit. A short is usually unintended and considered a failure, however conductors may be intentionally connected by a "shorting bar," as in a Data Connector plug.

Signal   The information conveyed through a communication system.

Signal-to-Noise Ratio   The ratio of received signal level to received noise level, expressed in dB. Abbreviated S/N. A higher S/N ratio indicates better channel performance. (See also ACR.)

Single Mode Fiber   A fiber wave guide in which only one mode will propagate. Single mode fiber has a very small core diameter, in some fibers approximately 8 μm. It allows signal transmission for long fiber distances with relatively high bandwidths and is generally driven with a laser diode. (See also multimode fiber.)

Single-ended   An unbalanced circuit or transmission line, such as a coaxial cable transmission line. (See also balanced line.)

Sinusoidal   A signal which varies over time in proportion to the sine of an angle. The description is often given to contrast with a nonsinusoidal waveform, which may contain significant harmonics above the fundamental frequency of the waveform. Ordinary alternating current is sinusoidal.

Skin Effect   The tendency of alternating current to travel on the surface of a conductor as the frequency increases.

S/N   Abbreviation for signal-to-noise ratio. (See also ACR.)

SNMP   Simple Network Management Protocol. A remote management protocol that is part of the TCP/IP suite of protocols. Devices that support SNMP can be monitored and controlled remotely over the network.

SNR   Abbreviation for signal-to-noise ratio; less common than the abbreviation S/N. (See also signal-to-noise ratio, ACR.)

SONET   Synchronous Optical Network. SONET is a fiberoptic technology that is used to transport telecommunications at 155 MBps and greater. It is the physical layer for several ATM implementations.

Source   In fiber optics, the device which converts the electrical information-carrying signal to an optical signal for transmission over an optical fiber. A fiber-optic source may be a light-emitting diode (LED) or a laser diode.

Speed of Light In a vacuum, 2.998×10^{-8} meters per second. Abbreviated c. See also, nominal velocity of propagation.

Split Pair A wiring error in twisted pair cabling where one of a pair's wires is interchanged with one of another pair's wires. Split pair conditions may be determined with a transmission test. Simple dc continuity testing will not reveal the error, because the correct pin-to-pin continuity exists between ends. However, the error may result in impedance mismatch, excessive crosstalk, susceptibility to interference, and signal radiation.

Step-Index Fiber An optical fiber in which the core is of a uniform refractive index with a sharp decrease in the index of refraction at the core-cladding interface.

Stitching The activity of terminating multiconductor cables on a punchdown block.

STP Shielded Twisted Pair. (See also UTP).

Structured Wiring Telecommunications cabling that is organized into a hierarchy of wiring termination and interconnection structures. The concept of structured wiring is used in the common standards from the TIA and EIA.

Surge A temporary increase in the voltage in an electric circuit or cable. Also called transient voltage.

T1 A specially framed 1.544 MBps wide area transport system which uses two twisted pairs for transport between a telephone central office and a customer premise. The data stream is channelized into 24 channels at 64 kBps. The system was originally designed for digitized voice transmission, but is often used to link digital data systems, such as LANs.

TCP/IP Transmission Control Protocol/Internet Protocol. A protocol used in the interconnection of computers. Although TCP/IP is actually a single protocol, the term is generally used to refer to a suite of closely related protocols for networking. The TCP/IP suite is used extensively in the Internet.

TDR Time Domain Reflectometry. A technique for measuring cable lengths by timing the period between a test pulse and the reflection of the pulse from an impedance discontinuity on the cable. The returned waveform reveals many undesired cable conditions, including shorts, opens, and transmission anomalies due to excessive bends or crushing. The length to any anomaly, including the unterminated cable end, may be computed from the relative time of the wave return and the nominal velocity of propagation of the pulse through the cable. (See also nominal velocity of propogation, speed of light.)

Teflon A DuPont Company trademark for fluorocarbon resins. (See FEP and TFE).

Terminator A resistor/connector combination that is connected to the end of a transmission line, such as a coaxial cable, to match the characteristic impedance of the cable. The terminator absorbs all signal energy that reaches it, thereby eliminating unwanted signal reflections from the end of the cable.

TFE Tetrafluoroethylene. A thermoplastic material with good electrical insulating properties, chemical resistance, and combustion resistance.

Thicknet Ethernet coaxial trunk cabling. (See 10Base5.)

Thinnet Ethernet thin coaxial cabling. (See 10Base2.)

TIA Telecommunications Industry Association. A membership association of manufacturers and users that establishes standards and publishes test methods. The body which authored the TIA/EIA 568-A "Commercial Building Telecommunications Wiring Standard" in conjunction with the EIA. (See also EIA.)

Tip A polarity designation of one wire of a pair indicating that the wire is that of the primary (common) color of a 5-pair group (eg., the white-blue wire of the blue pair); a wiring contact to which the tip wire is connected; the positive wiring polarity. (See also ring.)

Token-Ring A LAN topology and access method controlled by passing a digital token along a wiring ring formed between connected devices. Rings may operate at either 4 or 16 Mbps and device connections to the ring are made by means of a Multistation Attachment Unit (MSAU or MAU). Implemented originally by IBM, Token-Ring is standardized in IEEE/ISO 8802-5 (IEEE 802.5).

Topology The physical or logical interconnection pattern of a LAN.

Tracer The contrasting color-coding stripe along an insulated conductor of a wire pair.

Transmission Line An arrangement of two or more conductors that is used for signal transmission.

Transmitter A device that converts electrical signals for transmission to a distant point.

Twisted Pair A communication cable using one or more pairs of wires that are twisted together. When driven as a balanced line, the twisting reduces the susceptibility to external interference and the radiation of signal energy.

Type N Connector A threaded-barrel constant-impedance coaxial connector for large diameter cables.

UL Underwriters Laboratories, Inc.

Unbalanced Line A transmission line in which voltages on the two conductors are unequal with respect to ground. Generally one of the conductors is connected to a ground point. An example of an unbalanced line is a coaxial cable. (See also balanced line.)

Underground Cable Cable that is intended to be placed beneath the surface of the ground in ducts or conduit. Not necessarily intended for direct burial in the ground.

User Cord (See patchcord.)

USOC Universal Service Order Code.

UTP Unshielded Twisted Pair, a type of transmission cable.

V Symbol for volt.

Vampire Tap The method of connecting a transceiver to an Ethernet trunk cable by drilling an access hole in the cable, inserting a contact to the center conductor, and clamping the transceiver onto the cable at the tap.

Velocity of Propagation (See nominal velocity of propogation.)

VHF   Very high frequency. The portion of the electromagnetic spectrum extending from 30 to 300 MHz.

Video   A signal which contains visual information, such as a picture in a television system.

Volt   The unit of electrical potential. One volt is the electrical potential that will cause one ampere of current to flow through one ohm of resistance.

Voltage   Electrical potential or electromotive force expressed in volts.

Voltage Drop   The voltage developed across a component by the current flow through the resistance of the component.

W   Symbol for watt.

Watt   A unit of electrical power. One watt is equivalent to the power represented by one ampere of current flowing through a load with a voltage drop of one volt in a dc circuit.

Wave Form   The amplitude of a signal over time.

Wavelength   The physical distance between successive peaks or nodes of a wave in a transmission medium. Wavelength may be calculated from the frequency of a signal and the nominal velocity of propogation. In a vacuum, the NVP is simply the speed of light.

WAN   Wide Area Network. An extension of a local computer network that extends the connection beyond a location, usually with telecommunications common carriers.

Z   Symbol for impedance.

Index

ABOUT THE AUTHOR

James Trulove has more than 20 years of experience in
LAN and datacom systems with companies such as
Motorola ISG and Intel, and is currently affiliated with a
leading distributor of networking systems. He has extensive
experience in designing, installing, and troubleshooting
LAN cabling systems. He writes on local and wide area
networking topics and is the author of *A Guide to
Fractional T1*.